MAPPING BY DESIGN

A GUIDE TO ArcGIS MAPS FOR ADOBE CREATIVE CLOUD

MAPPING BY DESIGN

A GUIDE TO ArcGIS® MAPS FOR ADOBE® CREATIVE CLOUD®

SARAH BELL

Esri Press
REDLANDS | CALIFORNIA

Esri Press, 380 New York Street, Redlands, California 92373-8100
Copyright © 2021 Esri
All rights reserved.
Printed in the United States of America
25 24 23 22   2 3 4 5 6 7 8 9 10

ISBN: 9781589486041

Library of Congress Control Number: 2021943194

CONTENTS

PREFACE

OFTEN, CARTOGRAPHY—OR MAPMAKING—IS defined as a field at the intersection of science and art. Maps can illustrate scientifically collected and analyzed spatial data, and by investing energy into making them comprehensive and beautiful, you can make maps into artistic works. Every one of you is creative, and cartography requires technical skills, for which there is no circumventing. Whether you are using a pencil or a computer program, learning how to make a map takes effort. The great news is, it is fun! Every step along the learning journey will open new avenues for you to exercise your creativity. If you are new to making maps, here is a wonderful secret: cartography is an ever-evolving field with the promise of lifelong learning. Even we professional cartographers are still students of the craft. It is, in part, this feeling of perpetual newness that fuels my passion for cartography.

During this ongoing journey as a cartographer, I, like many of you, have relied on tutorials as a necessary ingredient for learning the craft. Because tutorials have played an important role in my learning how to use mapmaking tools, I wrote this book as a tutorial-centered guide for learning ArcGIS® Maps for Adobe® Creative Cloud® workflows.

Throughout this book, I use the terms *cartographer*, *map designer*, *mapmaker*, and *mapper*—and possibly a few others—to encompass one role: cartographer. It is my belief that, no matter where you are in your mapmaking career, the act of mapping is what makes a cartographer. A diverse range of cartographers is necessary for understanding the spatial stories that maps can uncover. We learn these stories by respecting maps from mapmakers new and experienced. I hope that this helps you, especially the newer cartographers, move forward with confidence and a desire to pursue the craft's skills. I also hope you all enjoy this mapping journey as much as I do.

Finally, I need to tell you how Maps for Adobe Creative Cloud began. When I started at Esri, Clint Loveman was my manager. Clint's energy was joyful and boundless. He put his full trust in my cartography and data visualization skills, which has allowed me to flourish professionally. Clint had the idea of connecting Creative Cloud to Esri's online geospatial platform, ArcGIS Online. So, he assembled a small team to accomplish this, a team that I am so grateful to have been on from the beginning. First, we made a prototype. It was quite rewarding to see what the team accomplished. Then the application grew into a released software available to the public. Today, Maps for Adobe Creative Cloud continually grows because of the fantastic team that Clint initiated. Maps for Adobe Creative Cloud—and this book—are possible because of him.

It is my hope that as Maps for Adobe Creative Cloud users, you are part of the continued evolution of this extension. I look forward to seeing your maps!

Sarah Bell
June 2021

ACKNOWLEDGMENTS

THE ArcGIS® MAPS FOR ADOBE® CREATIVE Cloud® extension is made possible by a team of talented people orchestrating its development daily. I simply cannot complete a book about Maps for Adobe Creative Cloud without thanking the many people who have contributed to its growth and development. I have immense gratitude for the user community, testers, conference-goers, coworkers, and friends who have provided feedback, watched demos, or just listened to my musings on mapping. You all have positively impacted this product. Following is a list of key individuals who have played important roles in Maps for Adobe Creative Cloud. As these lists go, it is imperfectly incomplete. I thank the following:

Valerie Alcantara, Javier Angel, Anna Breton, Jino Contreras, Jordan Cullen, Debalin Das, Ixshel Escamilla, Kara Goodman, Natalie Hansen, Elena Hartley, Jennifer Hill, Wendy Kallio, Aubri Kinghorn, Manuel Lopez, Clint Loveman, Nancy Morales, Elena Maklakova, Steven Moore, Joseph Munyao, Guy Noll, Kenneth O'Guinn, Rajeev Pahuja, Brandy Perkins, Madhura Phaterpekar, Siva Pidaparthi, Greg Pleiss, Salman Rafique, Alan Taylor, Vladimir Tkachov, Dmitry Travin, Rey Umali, Craig Williams, Jeremy Wright, Konstantin Yakovenko, Alan Zhang, and Xingdong Zhang.

Thanks to all the talented people at Esri Press as well.

And finally, thank you to my always supportive partner, Joseph Ferrare.

INTRODUCTION

THIS BOOK IS FOR MAPMAKERS FROM ALL backgrounds who want to learn ArcGIS® Maps for Adobe® Creative Cloud®. Mapmakers range from novices to skilled graphic designers to technical geographic information scientists to everyone in between. Because of the wonderfully diverse array of mapmakers, this book's audience has a wide range of skills. Some of you have spent a lot of time developing your own aesthetic style in graphic design, whereas others have years of experience in geographic information systems (GIS). Many of you bestride the worlds of artistic design and GIS in your daily work. Maps for Adobe Creative Cloud bridges GIS and graphic design by connecting Adobe® Illustrator® to Esri's ArcGIS system. It is this connection that makes Maps for Adobe Creative Cloud valuable to mappers from an extensive range of talents. Because of the breadth of knowledge that you—this book's readers—bring to mapping, this book seeks to provide the principles behind map design and map science. Several of these principles are demonstrated with maps, including award-winning maps from the first-ever ArcGIS Maps for Adobe competition, in 2020.

However, this book is ultimately meant as a guide for learning Maps for Adobe Creative Cloud. This type of guidebook cannot dive deep into GIS principles without risking a shift in focus from its main purpose: teaching you how to use Maps for Adobe Creative Cloud. Likewise, as this book is a Maps for Adobe Creative Cloud learning tool, it should not be considered an Illustrator guide, yet Illustrator tips are peppered within.

In this book, you will learn by doing. Chapter 1 provides what you need to know to get started using Maps for Adobe Creative Cloud, and then beginning in chapter 2, you will be using step-by-step tutorials that will teach you how to use this mapping extension by making maps. This experiential learning approach demonstrates many Maps for Adobe Creative Cloud workflows, instead of only presenting the software features as disparate parts. I recommend that during your first time through the tutorials, you follow them as written.

Although Maps for Adobe Creative Cloud can be used in both Illustrator and Adobe Photoshop®, this book's tutorials are written for the Illustrator workflow, as this is most common for the extension's users. As author, I use Adobe's term *extension* to refer to Maps for Adobe Creative Cloud. However, you could interchange *plug-in*, *add-on*, or *app*. To perform this book's tutorials, you will need (1) an active Adobe Creative Cloud account, (2) Illustrator software, and

(3) an ArcGIS Online account (GIS Professional or Creator license level), a Maps for Adobe Creative Cloud Plus account, or the trial license of ArcGIS software provided to purchasers of this book. Details for accessing this trial license are provided in chapter 1.

Maps for Adobe Creative Cloud is an extension that can be used to make maps directly in Illustrator or to open maps in Illustrator that were made using ArcGIS Pro. Chapters 2 through 6 guide you through how to use Maps for Adobe Creative Cloud to make maps, and chapter 7 details steps for ArcGIS Pro map and layout setup for the purposes of opening exported files in Illustrator via the extension. As you will discover, once your maps that originated in ArcGIS Pro are open in Illustrator, you can continually build upon them using Maps for Adobe Creative Cloud, thereby applying the skills presented in chapters 2 through 6.

ArcGIS Pro is not required to perform any of the tutorials in chapters 2 through 6.

Table 1 lists Maps for Adobe Creative Cloud keyboard shortcuts that can be used when the extension's panels are active. You can refer to this table when working with Maps for Adobe Creative Cloud in this book's tutorials and in the field. Illustrator provides many user-friendly keyboard shortcuts as well. When you hover your mouse pointer over several of the tool icons in Illustrator, a ToolTip appears showing their shortcuts.

In each tutorial, there is abundant space for creatives to imprint their unique aesthetic style while also having a solid mapmaking workflow framework to lean on. I am eager to see the creative maps that you produce from these tutorials.

Table 1. Maps for Adobe Creative Cloud keyboard shortcuts

Action	Windows	macOS
Move/Drag tool (hand tool)	Press and hold Spacebar	Press and hold Spacebar
Marquee Zoom In tool	Hold Ctrl + Spacebar	Hold Cmd + Spacebar
Marquee Zoom Out tool	Hold Ctrl + Alt + Spacebar	Hold Cmd + Spacebar + Option
Zoom In Once	Ctrl + Plus Sign	Cmd + Plus Sign
Zoom Out Once	Ctrl + Minus Sign	Cmd + Minus Sign
Fit To Screen*	Ctrl + 0	Cmd + 0
Zoom In/Out**	Ctrl + mouse wheel up/down	Cmd + mouse wheel up/down

*Compilation panel only

**On the Mapboards panel, pressing Ctrl (Cmd in macOS) is not necessary.

CHAPTER 1

WELCOME TO ArcGIS MAPS FOR ADOBE CREATIVE CLOUD

WHETHER YOU ARE A NEW MAPMAKER OR HAVE been a professional cartographer for decades, the mapping software you choose will play a significant role in your productivity and final map products. Many tools are available to today's map designers. Researching which of these tools to include in your workflow can be time consuming. Furthermore, you may already be accustomed to specific mapping software programs, and the journey of conquering a new one can feel unfamiliar and daunting. Your time is valuable, so spending it learning a new tool should benefit you by enhancing your skills and increasing efficiency, while also empowering you to make effective maps. The ArcGIS® Maps for Adobe® Creative Cloud® app (Maps for Adobe Creative Cloud henceforth), Esri's new mapping tool, has been developed specifically to provide map designers with a familiar experience by enabling you to create maps using software you already know.

Advantages of Maps for Adobe Creative Cloud

Maps for Adobe Creative Cloud is an extension that connects Adobe® Illustrator® and Adobe® Photoshop® to the power of ArcGIS technology, including ArcGIS Pro, Esri's desktop GIS application, and ArcGIS Online, Esri's software as a service and geospatial platform. ArcGIS Online provides a collection of map data and functionality that is accessed through a browser, allowing users to create maps and apps, share maps and data, and perform data analysis. By connecting Illustrator and Photoshop to ArcGIS Online, Maps for Adobe Creative Cloud gives designers the power to create maps by providing easy access to authoritative digital maps and map layers. In Maps for Adobe Creative Cloud, this seamless connection means that you can add these digital map layers, perform map enhancements and geo-analytical functions, and then download your maps as well-organized, ready-to-design files in Illustrator and Photoshop. For ArcGIS license users, these downloaded files are automatically synced to the Maps for Adobe Creative Cloud extension, a tremendously convenient feature for mapmakers who are designing multiple maps of the same area or who want to add new map layers and perform additional geo-analyses to their maps after download. You will learn more about the convenience of synced maps in chapter 3.

Chapter 1 shows how Maps for Adobe Creative Cloud provides access to a wide variety of map data, enabling creative mapmakers like you to design accurate and beautiful cartographic products. You will also learn about the different Maps for Adobe Creative Cloud licenses and the features provided in each one. This chapter concludes with information on how to install Maps for Adobe Creative Cloud and how to sign in to the extension.

Improving workflows

Adobe Illustrator is the world's leading vector graphic editor application. Because of Illustrator's abundant high-quality vector design tools and features, cartographers have long used this application as one of their go-to mapmaking tools. Two common approaches that cartographers have traditionally followed when incorporating Illustrator in their map projects are the geographic information system (GIS)–to–Illustrator workflow and the trace-and-place workflow. The GIS-to-Illustrator workflow is widely used by many professional cartographers, whereas the trace-and-place workflow is a common approach for many graphic designers and creatives.

GIS-to-Illustrator workflow

The GIS-to-Illustrator method requires uploading geographic data layers (vector, raster, or both) to a GIS application. Once the data are added to a GIS map project, and perhaps some geo-analyses are performed

in the GIS application, the next step is to export the map as a file type that can be opened in Illustrator. When the file is opened in Illustrator, the final step in this GIS-to-Illustrator workflow is the cleanup step, which is a series of many tedious steps. Cleaning up a map in Illustrator that has been exported from a GIS application can include layer organization and layer reordering, removing unwanted clipping masks and ungrouping artwork, joining broken lines, and fine-tuning text and labels to bring the map to a ready-to-design state. The actual aesthetic design process usually does not begin until the Illustrator file is ready to design. Depending on how complex and detailed the map is, the cleanup process can take several hours. Sometimes, in this workflow scenario, map labels are imported with each label's letter as an individual Illustrator text path, which can add even more time to map cleanup. With Maps for Adobe Creative Cloud, labels are complete, with all characters in a single string for easy text editing.

Trace-and-place workflow

The second approach that mapmakers have traditionally used to create maps with Illustrator is the time-consuming trace-and-place workflow. This method is performed by many designers who need to create beautiful maps but do not have an immediate need for a high-powered GIS application. By placing images, such as aerial and satellite imagery or photos of existing maps, into an Illustrator file, trace-and-placers trace individual map features such as roads, rivers, and boundaries, drawing each individual element as digital artwork with Illustrator tools such as the Pen or Pencil tool. Points of interest and other map features are also added by using an array of tools available in Illustrator. Labels and text are manually typed one by one. The arduous workflow of creating each map feature, anchor point by anchor point and label by label, can be laborious, leaving map designers longing for a more efficient method.

We cartographers are a diligent bunch. Our final products are not just graphic design; when we make maps, we are also conveying visual-spatial stories that must be well understood by our intended audiences. We do all of this while also aiming for stunning aesthetics. We know that providing meaningful and accurate maps to our clients requires us to be meticulous. For this reason, many mapmakers have accepted that the time-consuming tasks in the GIS-to-Illustrator and trace-and-place workflows are just part of the job—as though they were required to get a map to its design-ready state. But what if you could get to the aesthetic design phase of your mapping projects much sooner, and with access to accurate, detailed map layers and powerful mapping tools that could enhance the map's story? In the following section, you will read about the ways that Maps for Adobe Creative Cloud can help eliminate many of these time-consuming undertakings for mapmakers like you.

Avenues for efficient organized map production

Just as there are two traditional approaches for incorporating Illustrator into map design workflows, there are two corresponding avenues for map creation using Maps for Adobe Creative Cloud. Many of you—like me—will find yourself taking advantage of both. As you read this book and follow its tutorials, you will also discover that these two avenues can have significant overlap in many mapmaking steps

and procedures. The main difference between the two Maps for Adobe Creative Cloud approaches is where you begin your mapmaking process.

Extension-direct workflow: Creating directly from Maps for Adobe Creative Cloud

One avenue to integrate Maps for Adobe Creative Cloud into your mapmaking workflow is to create your maps directly with the extension open in Illustrator (figure 1.1). This seamless process appeals to creative designers as well as GIS cartographers who want access to copious map data and layers but do not need a desktop GIS application to produce a map. Even with this lightweight approach, the Maps for Adobe Creative Cloud extension includes many ArcGIS geo-analytical and data visualization tools and processes. Mapmakers can create highly detailed cartographic products—from start to finish—by beginning their maps directly from Maps for Adobe Creative Cloud. Because high-quality detailed maps are possible with this extension-direct workflow, creating maps directly from Maps for Adobe Creative Cloud offers considerable benefits to cartographers who have traditionally followed the trace-and-place workflow, and often offers adequate features for mappers who are accustomed to the GIS-to-Illustrator workflow as well.

For the trace-and-placers, Maps for Adobe Creative Cloud replaces the need to individually draw or place map features one by one. Now, users can browse an abundance of map and data layers, including popular city-, regional-, country-, and world-level data that can be added to a map with the click of a button. Maps for Adobe Creative Cloud makes it easy to add points of interest, transportation, administration boundaries, and geophysical data to a map workspace

Figure 1.1. *Top American Donut Shops* is one of the first maps I made using the Maps for Adobe Creative Cloud extension-direct workflow. The map features the top 31 donut shops listed in a September 2018 Thrillist.com article, "The 31 Best Donut Shops in America."

that can be downloaded as an Adobe Illustrator (AI) file with editable digital artwork (figure 1.2). Other time-saving options include adding map labels, filtering which features you want to display, and a host of various geo-visualization and geo-analysis operations. These easy-to-access features make the Maps for Adobe Creative Cloud extension-direct workflow an efficient solution for cartographers from all backgrounds.

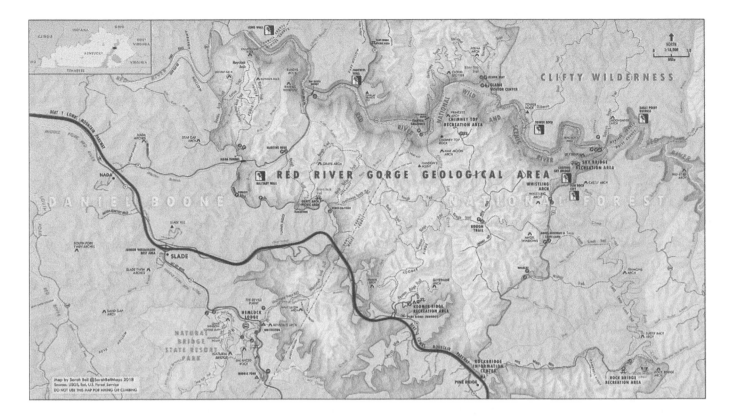

Figure 1.2. This *Red River Gorge Geological Area* map features a large portion of the Red River Gorge in Kentucky. The area includes hiking and world-class climbing on the gorge's sandstone cliffs. As a climber and mapper, I often enjoy creating maps that feature some of my favorite climbing spots. This map was also created using the extension-direct workflow. I used a combination of data layers that were available through both ArcGIS Online and local computer files. Later in this book, you will learn how to add data from both sources.

ArcGIS Pro-to-Illustrator

The second Maps for Adobe Creative Cloud workflow that you can use is the ArcGIS Pro–to–Illustrator workflow. Map creation for this workflow begins in ArcGIS Pro. This workflow is ideal for cartographers who rely on the robust GIS toolbox of ArcGIS Pro tools and processes. When the ArcGIS Pro map or layout is set the way you want it—with the appropriate coordinate systems, required map layers, labels, geo-analyses, and map elements—you can export the map or layout from ArcGIS Pro as an Adobe Illustrator Exchange (AIX) file. When signed into Maps for Adobe Creative Cloud, you can open AIX files in Illustrator. Note that to open an AIX file, you must have an ArcGIS Online or ArcGIS Enterprise license. Maps for Adobe Creative Cloud processes the AIX file into a well-organized, ready-to-design map in the form of an AI file. This ArcGIS Pro–to–Illustrator approach provides mapmakers who need the powerful, high-end geo-analytical capabilities of ArcGIS Pro the opportunity to complete their map's aesthetic design in a familiar graphics editor.

After the AIX file is opened in Illustrator, the converted AI file layer structure matches the layer hierarchy of the source ArcGIS Pro project. Map label arrangement is also maintained in both placement and congruity, making labels easily editable, and map features' styles and appearances are rendered as editable Illustrator artwork. Chapter 7 further details the Arc-GIS Pro–to–Illustrator workflow.

Software and licenses

This section outlines the required software and licensing options for using this book. Choose the licensing option that is best for you.

Adobe

At the time of this book's publication, to use Maps for Adobe Creative Cloud, your Adobe Creative Cloud subscription must have Illustrator version 24.0 (2020) or later installed for the extension to work properly. Licenses and installation instructions for these products should be obtained from Adobe. Please see links.esri.com/MapsForAdobeResources for more information on supported versions.

Maps for Adobe Creative Cloud

Users can license Maps for Adobe Creative Cloud by choosing either an ArcGIS Online, Plus, or Complimentary license. Features and functionality for each license vary, as do their commercial usage allowances (table 1.1). See also links.esri.com/FunctionalityMatrix. For up-to-date information, see "Install Maps for Adobe Creative Cloud" in the product documentation at links.esri.com/AboutMapsForAdobe.

ArcGIS Online

ArcGIS Online organizations can assign different user types granting specific privileges. You must have an ArcGIS Online account of one of the following user types: Creator, GIS Professional, Editor, or Viewer. Although these four user types all have access, at the time of this book's publication, the extension's functionality varies. Creator and GIS Professional user types are granted full access to all functionality, whereas Editor and Viewer user types cannot run geospatial analyses or save their Maps for Adobe Creative Cloud compilations to their ArcGIS Online account. **Note:** Several of the tutorials in this book include at least one step that requires a Creator or GIS Professional user type, so one of these user types will be necessary to complete those tutorials. See "User types, roles, and privileges" in the ArcGIS Online help at links.esri.com/UserTypes for more information about ArcGIS Online user types.

Use an existing ArcGIS Online or ArcGIS Enterprise license

If you already have credentials for a Creator, GIS Professional, Editor, or Viewer user type, you can sign in to Maps for Adobe Creative Cloud and extend your ArcGIS privileges with Maps for Adobe Creative Cloud features and content. This license also allows you to open AIX files in Illustrator that were created using ArcGIS Pro or ArcGIS Online Map Viewer.

Table 1.1. Functionality matrix

Function	Complimentary	Plus	ArcGIS (Viewer, Editor)	ArcGIS (Creator, GIS Professional)
Content access				
Import extent and data from no-cost public content	✓	✓	✓	✓
Import extent and data from subscriber content and premium content		✓	✓	✓
Plot addresses and map data in bulk from local files (geocode)[1]	✓	✓	✓	✓
Search and plot locations based on place-name or keyword[2]	✓	✓	✓	✓
Automated graphic processes (legends, custom symbols, and more)	✓	✓	✓	✓
Access to Maps for Adobe Creative Cloud content	✓	✓	✓	✓
Access many of Esri's vector tile basemaps and layers			✓	✓
Visualize travel times and routes, create buffers, enrich layers with demographic attributes[3]		✓		✓
Advanced features				
Mark a layer or web map as a favorite		✓	✓	✓
Save the map as part of an Illustrator or Photoshop document for future session modifications		✓	✓	✓
Use higher-resolution PPI for improved vector or raster data		✓	✓	✓
Apply custom projections and coordinate systems		✓	✓	✓
Open AIX files exported from ArcGIS Pro with the option to add more data			✓	✓
Share a map to your ArcGIS organization				✓
Usage				
Commercial and revenue-generating usage		✓	✓	✓

1 Number of locations that can be geocoded via a local shapefile, CSV, TXT, or KML file format is limited to 1,000 features for Complimentary users and 4,000 features for Plus and ArcGIS accounts.

2 Complimentary users only see the top 50 results when geosearching places.

3 Running these tools with an ArcGIS Online organizational account requires credits.

Use the 180-day trial account that comes with the purchase of this book

This book comes with a 180-day trial license of ArcGIS software. If you do not have a Creator or GIS Professional license, it is strongly recommended that you initiate the trial license when you are ready to begin this book's tutorials. The EVA code can be found inside the back cover of the print book. E-books rented or purchased through VitalSource come with an EVA code. Once you have located the EVA code, activate your license at links.esri.com/EVAcode. This trial will provide a license for the Creator user type. At the end of your trial, you can purchase an ArcGIS Online subscription, and all the work you have saved during the trial period will become part of your account.

ArcGIS for Personal Use subscription

ArcGIS for Personal Use comes with everything you need to access the full range of ArcGIS capabilities to create rich, dynamic maps and apps. This license provides the Creator user type, giving full access to Maps for Adobe Creative Cloud functionality.

Plus

This subscription-based license gives access to the extension's premium functionality at a low monthly cost. Features include premium maps and layers, the ability to sync downloaded files to the extension for future modifications, map projections, and other advanced features. In addition, the Plus license allows you to create and design maps for commercial and revenue-generating purposes. See "Terms of use and data attribution" in the product documentation for the full terms of use.

Complimentary

This license provides limited access to Maps for Adobe Creative Cloud features, including public maps and map layers hosted in ArcGIS Online. Sign up for this free-access license using an email address as your credential identification, and you will receive an access code. This license offers no time-out limits on a single machine. The Complimentary license cannot be used for commercial or revenue-generated purposes.

ArcGIS Enterprise

With Maps for Adobe Creative Cloud version 3.0 and later, Maps for Adobe Creative Cloud will be available to ArcGIS Enterprise users who are using ArcGIS 10.9 and later. Access to data hosted online via Maps for Adobe Creative Cloud requires an internet connection. For information about using Maps for Adobe Creative Cloud with an ArcGIS Enterprise license, see "Install Maps for Adobe Creative Cloud" in the product documentation, at links.esri.com/MapsforAdobe.

The tutorials in this book were designed for an internet-connected experience. For this reason, it is recommended that you do the tutorials with your ArcGIS Online Creator or GIS Professional license or the 180-day trial offered with purchase of this book.

Getting started with Maps for Adobe Creative Cloud

Download the software

1. Download Maps for Adobe from the product page, at links.esri.com/MapsForAdobe, and choose either the Microsoft Windows or MacOS version.

2. Install Maps for Adobe on your machine.

Open the extension

After installing Maps for Adobe Creative Cloud, you will find it listed in the Illustrator Extensions flyout list.

1. Open Illustrator.

2. On the top menu, go to Window > Extensions, and click ArcGIS Maps for Adobe Creative Cloud.

When Maps for Adobe Creative Cloud opens, the first thing you will see is the Sign In window with the corresponding options for each license (figure 1.3).

Figure 1.3. The ArcGIS Maps for Adobe Creative Cloud sign-in window appears when you open the extension in Illustrator.

Set up your account

Once you have determined which license is right for you, complete the remaining steps for setting up your account.

ArcGIS account sign-in

If you already have an ArcGIS Online or ArcGIS Enterprise account, no additional steps are needed to set up Maps for Adobe Creative Cloud once it is installed on your machine. You will be able to open Maps for Adobe Creative Cloud from the Illustrator Window > Extensions menu, and sign in using your ArcGIS Online account credentials.

1. Click the ArcGIS Online option, and enter your account credentials or your ArcGIS organization's URL to sign in to the extension.

Plus account setup and sign-in

1 Click Sign Up in the window, and then click the Plus option in the sign-up window to initiate your account setup. Next, you will complete a simple five-step form to supply the required information and create your sign-in credentials for the Plus license. Once the setup is complete, you can sign in to Maps for Adobe Creative Cloud.

Complimentary account setup and sign-in

1 Click Sign Up in the Sign In window, and click the Complimentary option in the Sign In window. When prompted, complete the Complimentary account form and verification to set up your new account. If you choose to upgrade to a Plus or ArcGIS Online account, sign out of the Complimentary account, and then sign in with your new license.

When you sign in, the Mapboards panel opens. Maps for Adobe Creative Cloud has three main components, called panels. See chapter 2 to learn about these components.

Mapping Helsinki

By adding two map frames to a layout in ArcGIS Pro, cartographer Anna Breton was able to export the inset and main maps as one single Illustrator-ready file. Chapter 7 details the multiple map frames workflow.

Of her map, Anna says, "Suomenlinna in Helsinki was a place I loved to visit growing up in Finland and is a popular tourist destination in the city. The 18th-century sea fortress allows visitors to explore tunnels, old structures, and the beautiful natural landscape."

Figure 1.4. The *Suomenlinna* map by cartographer Anna Breton was created using the ArcGIS Pro-to-Illustrator method.

CHAPTER 2

EXPLORING THE MAPS FOR ADOBE CREATIVE CLOUD USER INTERFACE

THIS CHAPTER BEGINS WITH A TOUR OF THE MAPS for Adobe Creative Cloud user interface. You will explore the three Maps for Adobe Creative Cloud panels: the Mapboards panel, the Compilation panel, and the Processes panel. The first two panels have corresponding sections in this chapter that both conclude with a brief tutorial in which you will create a map using the Maps for Adobe Creative Cloud extension-direct workflow. You will explore the Processes panel further in chapters 3 and 6.

Maps for Adobe Creative Cloud panels

Each of the three Maps for Adobe Creative Cloud panels corresponds to logical phases in the mapmaking process.

1. **Mapboards panel.** This panel is used to choose and define your map area's extent. An extent is the area on Earth that you want your map to show. In the Maps for Adobe Creative Cloud extension-direct workflow, the extent is defined on the Mapboards panel. In the ArcGIS Pro–to–Illustrator workflow, the map extent is defined in ArcGIS Pro by either the map frame or the map. When an Adobe Illustrator Exchange, or AIX, file is opened in Illustrator, the extent automatically loads as a mapboard.

2. **Compilation panel.** This panel is used to search for and add map layers and to perform optional map customizations, such as adding labels, changing a map projection, and doing geo-analyses. These map-building actions and customizations can be performed on maps that are built directly within the extension as well as on maps that were created using the ArcGIS Pro–to–Illustrator workflow. See the functionality matrix in table 1.1 in chapter 1 for more on how features available on the Compilation panel vary by license type.

3. **Processes panel.** Many cartographic enhancements made during and after the map-building workflow are set up on the Processes panel. These processes can be performed on maps created using either the Maps for Adobe Creative Cloud extension-direct or ArcGIS Pro–to–Illustrator workflows.

What is a mapboard?

One of the new terms you will learn as you dive into Maps for Adobe Creative Cloud is *mapboard*. Mapboards are a vital component of this extension's map creation process. This section outlines some important things to know about mapboards. For AIX files exported from ArcGIS Pro, the mapboard spatial extent, size, name, and scale are determined by settings in ArcGIS Pro. See chapter 7 for more details on mapboards created from AIX files.

Mapboards and spatial extent

Simply put, mapboards represent the area of the Earth that you are mapping. In the extension-direct workflow, mapboards can be drawn manually, or they can be created dynamically by allowing the extent of a map layer's geographic features to define the mapboard. But a mapboard does much more than simply define the map's spatial extent.

Mapboards and size

In addition to determining your map's spatial extent, a mapboard defines the size of the map's final output.

For example, if you want to make a map that is 1,920 × 1,080 pixels, you will indicate this in the Mapboard Options dialog box (see the "A. Mapboards panel toolbar" section in this chapter). By doing so, when you download your map from Maps for Adobe Creative Cloud, an Adobe Illustrator, or AI, file with an artboard size of 1,920 × 1,080 pixels will be created.

Mapboards and map names

A mapboard also requires a name. If you do not specify a mapboard name, Maps for Adobe Creative Cloud will provide a default name that begins with the word "Mapboard." Tutorial 2.1 will demonstrate how to customize a mapboard name. Once you download your map as an AI file, the name of your downloaded AI file will be identical to the name of your mapboard. However, if you download a map to a directory that already has an AI file with the same name, Maps for Adobe Creative Cloud will append a serialized number to the end of the AI file's name (for example, "Mapboard_1," "Mapboard_2," "Mapboard_3," and so on).

Mapboards and map scale

When you create your mapboard manually, the map scale is automatically calculated. By downloading the map, you are syncing it to Maps for Adobe Creative Cloud. Before a mapboard is synced, you can edit the map scale to fit your map's requirements. As you update the map scale, the map's final output size will also change, so make sure to check the final output size before downloading the map. See "Cartography corner: Map scale" for more information on the inextricable link between a map's size, extent, and scale.

Map scale

One of the defining characteristics of maps is that they are abstractions. A map cannot depict all things occurring within its extent, nor should you want it to. Even if it were possible to show all spatial phenomena, from blades of grass to houses to city boundaries, the map would be a busy, un-readable mess. As the cartographer, you have the responsibility to choose which details to include and which to omit, so that your map's story is more readily conveyed.

One unavoidable abstraction that is included in all maps is scale. In cartography, scale is a ratio that describes how big or how small a map's represen-tation of objects is in relation to the actual objects it represents. For example, a map scale (ratio) of 1:100,000 means that one unit of measure-ment on the map equals 100,000 of those units of measurement on the ground. When a map's scale is expressed as a ratio, you can apply any unit you want to the representative fraction. So, on a 1:100,000-scale map, one inch equals 100,000 inches, and one centimeter equals 100,000 cen-timeters, and so on. Map scale can also be repre-sented by a scale bar, which might show a one-inch bar equaling 10 miles, for example.

As you zoom in, or make your map scale (ratio) larger, the map's representational objects will become larger. When you zoom out, the scale becomes smaller, and the map's representational objects become smaller. As mentioned in "What is a mapboard?," changing the mapboard scale—while keeping the spatial coverage

constant—will change the size of your map. Figure 2.1 helps illustrate the inextricable relationship between spatial coverage (extent), map size, and map scale. Imagine that both rectangles in the figure represent maps with identical spatial extent—they cover the same area of the Earth. If you want to keep the left map's same spatial extent while changing its scale from 1:100,000 to a larger scale (more zoomed in) of 1:10,000, it will increase the map objects' size 10 times. Therefore, the map will also be 10 times larger. The right map illustrates this increase in size and scale.

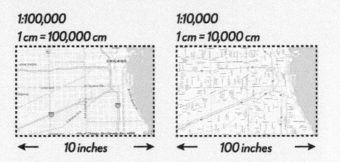

Figure 2.1. Two rectangles with identical geographic ex-tent. To retain this same extent while changing the scale from 1:100,000, *left*, to 1:10,000, the map on the right will increase in size by 10 times.

This does not mean that all your large-scale maps must be enormous. Typically, large-scale maps cover less area than small-scale maps. For example, 1:100,000 maps are useful at a regional level, such as for showing a city. Maps at a 1:10,000 scale are more appropriate for the neighborhood or city block level.

The following two maps were made with Maps for Adobe Creative Cloud using the extension-direct workflow. These two maps have different scales, even though their final output sizes are similar. This allows both maps to fit on the pages in this book. The first is a large-scale (1:6,000) map of downtown Seattle, Washington (figure 2.2), whereas the next is a smaller-scale map (1:96,000) focused on the ferry routes of Victoria, British Columbia, past and present, over the Strait of Juan de Fuca (figure 2.3).

Figure 2.2. *Seattle* by Sarah Bell. *Map data sources: OpenStreetMap; Esri.*

Figure 2.3. *Victoria Ferry Routes—Past and Present* by Sarah Bell. *Map data source: Esri.*

Syncing details: ArcGIS Pro-generated mapboards

A mapboard that is automatically loaded into the Maps for Adobe Creative Cloud extension when an ArcGIS Pro-generated AIX file opens is considered synced to Maps for Adobe Creative Cloud; therefore, the mapboard's name, size, extent, and scale are fixed and unchangeable.

Mapboards panel features

After you sign in to Maps for Adobe Creative Cloud, the Mapboards panel is the first component you will see (figure 2.4). The Mapboards panel is where a mapboard extent is defined when following the extension-direct workflow. Following are the Mapboards panel's elements, listed in the figure as A–J.

A. Mapboards panel toolbar

The initial state of the Mapboards panel toolbar contains the tools and parameter settings for mapboard creation. Once a mapboard is created, the toolbar will show more contextual options (figure 2.5), which can be used to further refine and toggle between multiple mapboards.

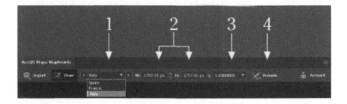

Figure 2.5. When a mapboard is drawn or selected, the Mapboards panel toolbar updates to reveal additional contextual mapboard setting options, 1–4.

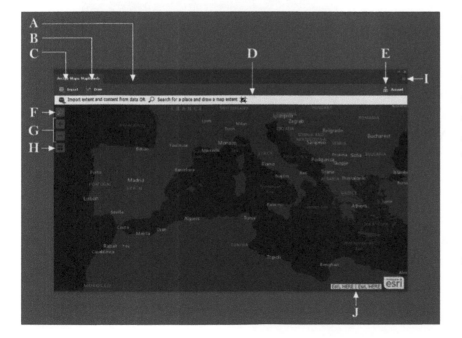

Figure 2.4. The Mapboards panel is the first component to open when you sign in to Maps for Adobe Creative Cloud. This is where the spatial extent of the map is defined and viewed. When an AIX file created from ArcGIS Pro is opened, the map extent can also be viewed on the Mapboards panel, although an AIX file-generated mapboard's name, size, extent, and scale cannot be edited.

1. Mapboard selector

When working with more than one mapboard during a single session, you can switch between mapboards by selecting the mapboard name from this drop-down list. You can also activate a mapboard by clicking the mapboard itself from the basemap area.

If an AIX file is generated from an ArcGIS Pro layout with multiple map frames, a mapboard will conveniently be created for each map frame.

2. Width and Height input fields

After creating a mapboard, you can adjust the final map size with the Width and Height input fields if the mapboard has not yet been synced. To lock the proportions of the mapboard width and height, click the Constrain Proportions button ⬚ located between the Width and Height text fields. A mapboard's width and height can also be adjusted on the Mapboard Options dialog box.

3. Scale selector

The map scale can be updated by either manually entering a value in the scale selector input area or by choosing one of the common ArcGIS Online map scales from the scale selector's drop-down list. You can also adjust the mapboard scale in the Mapboard Options dialog box (see tutorial 2.1).

4. Preview button

The Preview button opens the Compilation panel, where a preview of your map can be viewed. This preview updates as map layers and other map enhancements are added.

B. Draw mapboard tool

You can use the Draw mapboard tool to manually draw a mapboard. Clicking the Draw mapboard tool to turn it on will change the pointer to a marquee icon while it is within the basemap area. To draw a mapboard, drag the pointer from one spot on the basemap to another, making a box around the desired map extent. Release to finish drawing the mapboard. Once a mapboard is drawn, the Mapboard Options dialog box appears (figure 2.6) with options to further define the parameters of a mapboard. To disable the Draw mapboard tool, click the tool button once again.

Figure 2.6. Mapboard Options dialog box.

Map navigation

While the Draw mapboard tool is enabled, some of the map navigation features will become disabled, such as mouse wheel zooming capabilities and keyboard shortcuts. For an optimized mapboard creation experience, pan and zoom the basemap to your desired map location before enabling the Draw mapboard tool.

C. Import button

The Import button allows users to define their mapboard extent by the geographical extent of an ArcGIS Online data layer or web map or a local geo-data file.

When a layer or web map is added using the Import button, a mapboard is automatically created. The mapboard extent will match the geographic extent of the imported data. Clicking the Import button shows three options for adding map data, detailed in the following points. If the spatial extent of the imported map data differs from your desired spatial extent, you can manually adjust the mapboard after it is created.

Import options

- **Import From Layer.** This option allows you to add a map data layer from ArcGIS Online and to define the mapboard from that data layer.
- **Import From Web Map.** This option creates a mapboard encompassing the extent of an ArcGIS Online–hosted web map's most recently saved extent and scale. See "Overwrite From Map" in this chapter for more information.

- **Import From File.** With this option, you can define your mapboard area by the spatial extent of a local geo-file. Supported file types include shapefiles (SHP), comma-separated values (CSV), text files (TXT), Global Positioning System exchange (GPX), and keyhole markup language (KML and KMZ) files. Read about working with these file types in the "Local data" section in chapter 4.

Determining a map's extent

In the Maps for Adobe Creative Cloud extension-direct workflow, the most common way to create a mapboard is by manually drawing one. If you do not know the geographic extent of a map data layer or do not want to use a layer's geographic extent to define your map area, you can always begin by creating your mapboard with the Draw mapboard tool. After drawing a mapboard of the area you want your map to show, you will have the option to add a web map and map data layers on the Compilation panel.

D. Action bar

The Action bar is the yellow bar just below the Mapboards panel toolbar. When a mapboard is created, the Action bar updates with logical next-step prompts and buttons. To hide the Action bar, go to the Mapboards panel Settings (figure 2.7), and uncheck the Display Action Bar Tooltips box on the General tab.

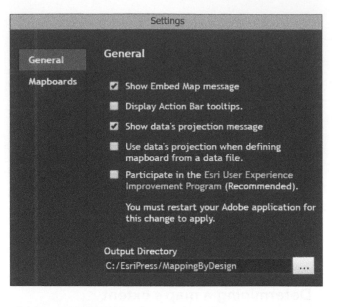

Figure 2.7. Mapboards panel settings.

E. Account button

When signing in to Maps for Adobe Creative Cloud with an ArcGIS Online or Plus account, you will remain signed in for two weeks or until you manually sign out.

F. Search tool

Click the Search tool to open the Search text field, and type a place-name, such as city, county, state, or address. This will center the map preview area on the searched location. A pushpin will appear on the basemap at the center of the searched location. This pushpin will not appear in your final map. You can remove the pushpin from the basemap area by clicking the x next to the location name in the Search tool text field.

G. Zoom In and Zoom Out buttons

Just as their names indicate, the Zoom In and Zoom Out buttons can be used to zoom into and out of the map. You can also zoom in or out with the corresponding keyboard shortcuts described in the introduction to this book.

H. Basemap selector button

The Esri Dark Gray Canvas basemap is a minimalist basemap that suits the default user interface settings in Illustrator, making it a good default basemap for Maps for Adobe Creative Cloud in terms of aesthetics. However, there will be times when you want more reference detail in the basemap when drawing a mapboard. You can choose from several basemaps by clicking the basemap selector button and selecting the basemap you want to use. If you select a map that is not available in the current mapping profile, an alert will appear next to the basemap's name, and you will be able to switch mapping profiles. Note that, at the time of this book's publication, the capability to switch mapping profiles is available only to ArcGIS Online and ArcGIS Enterprise users.

I. Mapboards panel application settings

From the Mapboards panel application settings, you can check for software updates, locate Maps for Adobe Creative Cloud help information, update the Mapping Profile application settings, and view and update the extension settings options. The Mapboards panel settings contains two tabs, the General and Mapboard tabs.

General

The General settings allow you to turn the Embed Map message (see tutorial 2.2) and Action bar on and off and to select whether to participate in the Esri User Improvement program. The Output Directory Path setting is also on this tab. This setting determines the location where the AI files will be saved when you download a map. You can update this path at any time.

Mapboards

The Mapboards tab can be used to show or hide the map scale description, which is found on the side toolbar and the Mapboards panel toolbar.

Mapping Profile

ArcGIS Online and ArcGIS Enterprise users can choose to create maps using the Legacy mapping profile or the AIX file mapping profile. AIX file mapping profiles support AIX files and vector tile basemaps and layers.

J. Map data sources

You can find the data sources for the mapboard's selected basemap here. If you select a new basemap, the map data sources will reflect the change.

Tutorial 2.1: Creating a mapboard

Scenario

This tutorial is the first in a two-part exercise. You will be creating a mapboard for the state of Minnesota using the extension-direct workflow. Your final product will be a letter-size map (8.5 × 11 inches) of Minnesota counties. You will complete this map in tutorial 2.2. In chapter 3, you will use the map you completed in tutorial 2.2 to create an enhanced version of the Minnesota map. Make sure to keep track of this map along the way, as you will be working with it for the first three tutorials of this book.

Learning objectives

- **Map navigation on the Mapboards panel:** Navigate the basemap using the Search and Zoom tools
- **Basemap selection:** Switch basemaps on the Mapboards panel
- **Mapboard creation:** Manually create a mapboard using the Draw mapboard tool
- **Mapboard customization:** Define mapboard parameters, such as name, size, and scale
- **Mapboard positioning:** Manually adjust the mapboard position using the mouse

Center the basemap over the desired location

1 Open Illustrator, and sign in to Maps for Adobe Creative Cloud using either your Plus or ArcGIS Online license or the free software trial available with this book.

After signing in, the Mapboards panel basemap position will be centered on a location determined by the account used to sign in. You are going to use the Search tool to reposition the basemap over Minnesota.

Figure 2.8. Searching for Minnesota with the Search tool will yield multiple results.

2 Click the Search tool and type **Minnesota** in the Search text field, and press Enter or Return.

3 Select the first option, Minnesota, USA (figure 2.8), to reposition the map in that state's location.

A pushpin is added in the geographic center of Minnesota.

4 Remove the pushpin by clicking *x* next to the word Minnesota in the Search field.

Select a new basemap with the basemap selector

1 Use the basemap selector (figure 2.9) to switch the basemap to a basemap of your choice.

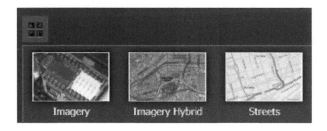

Figure 2.9. The basemap options may vary depending on the account you used to sign in.

2 Manually resize the Mapboards panel so that the entire state of Minnesota is visible. You can resize the panel by dragging the lower-right corner or the bottom and right edges of the panel. You may also want to zoom the basemap in or out. You can do this by clicking the Zoom In or Zoom Out tools or by using your mouse wheel.

MAPMAKER TIP

Resizing panels

Did you know that Illustrator panels can be manually resized and collapsed? To resize a panel, hover over the left, bottom, or right edge of the panel until the pointer switches to the resize icon. Then drag the panel's edge in the direction that you want to expand or shrink the panel. You can collapse a panel by double-clicking its name on the tab. To expand the panel again, click once on the tab.

Mapboard creation

Draw the mapboard

1 Click Draw to activate the Draw mapboard tool.

2 Place the pointer in the general spot marked by the letter *A*, and drag it to the general spot marked by the letter *B* (figure 2.10). Once the entire state of Minnesota is within your map extent, release to create the mapboard.

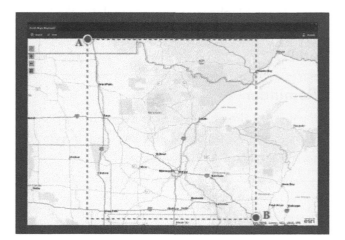

Figure 2.10. The Mapboards panel is focused on the state of Minnesota and has been resized to show the entire state. By dragging the pointer from approximately point *A* to point *B*, you will create a mapboard over this entire area.

Name the mapboard, and specify its size and scale

When you draw a mapboard, the Mapboard Options dialog box immediately opens, prompting you to provide a name along with other mapboard parameter options. The name that you provide will also be the name of the AI file that is created when you download your map.

 On the Mapboard Options panel, for the Name field, type **Minnesota**.

> **Reminder**
>
> If you have already closed the dialog box, open it again by clicking the Mapboard Options button on the Mapboards toolbar.

2 Under Set Artboard Size, click the down arrow next to Preset.

3 Expand Print, and select Letter from the list.

The preset units in your drop-down list are determined by the unit of measurement settings in Illustrator. When you choose Letter from this list, although the units may differ when converted to inches, it is the same as 8.5 × 11 inches.

4 Check that your scale is the same as in figure 2.11 (1:4,322,108). If it is not, click the scale's number directly, which allows you to manually type the precise scale. Click OK.

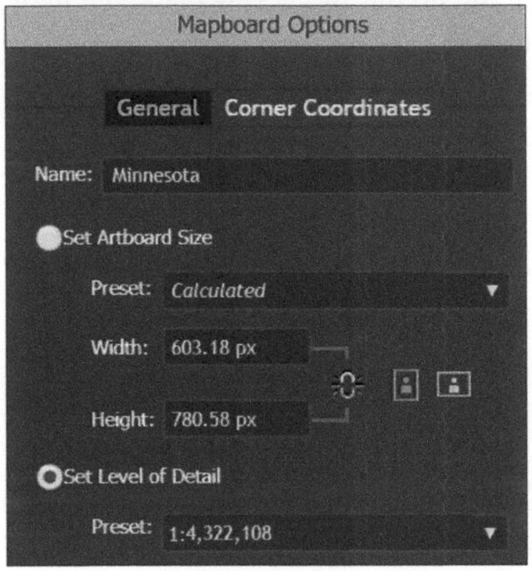

Figure 2.11. The Minnesota mapboard settings are updated in the Mapboard Options dialog box.

MAPMAKER TIP

Defining units of measurement
You can update the Illustrator unit of measurement setting by going to File > Document Setup. Then choose the desired unit of measurement. You will need to close and reopen Illustrator so that Maps for Adobe Creative Cloud recognizes these new unit settings.

Adjust the mapboard position

In the previous step, you updated a mapboard using custom parameters. Because these parameters were slightly different from the area that you drew under "Draw the mapboard," the mapboard's position may have shifted a little.

1 Check to see if your mapboard is still covering Minnesota.

2 If you need to move the mapboard, drag the mapboard's blue tab so that the entire state of Minnesota is within the mapboard.

Checkpoint

Nice work! You have just completed the first tutorial in this book. You made your first mapboard, provided it with a custom name, and updated the mapboard parameters to fit your project goals. On the Mapboards panel, you should have a mapboard that covers Minnesota (figure 2.12).

You also learned how to move the mapboard position by dragging its tab. You will be using your Minnesota mapboard in the next tutorial. If you need to close Illustrator before beginning the next tutorial, you will need to repeat the steps in tutorial 2.1 before moving on to tutorial 2.2.

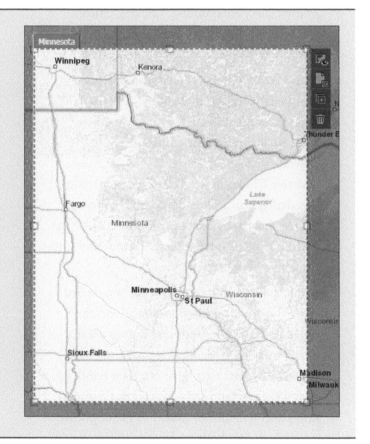

Figure 2.12. The Minnesota mapboard after it is created and the parameters have been applied. Your extent may differ slightly, which is fine if the entire state of Minnesota is within your mapboard extent.

Mapboard toolbar

The Mapboard toolbar, on the right side of the mapboard, is similar to the Mapboards panel toolbar and contains some of the same features, as well as additional features. The Mapboard toolbar contains the following buttons from top to bottom:

- **Preview:** This Preview button functions the same as the other Preview button on the Mapboards panel toolbar. Clicking either instance of the Preview button will open the Compilation panel.

- **Mapboard Options:** This button opens the Mapboard Options dialog box. A mapboard must be selected to enable this dialog box. A mapboard cannot be edited after it is downloaded and synced to Maps for Adobe Creative Cloud, in which case the settings in this dialog box will be dimmed. The settings for mapboards generated from ArcGIS Pro via AIX files will also be dimmed because they are considered synced.

- **Duplicate:** You can duplicate any mapboard created from the extension-direct workflow, synced or not. The Duplicate button creates a new mapboard identical in scale, size, and extent. Any data that have not been removed from the original mapboard on the Compilation panel will also be added to the duplicate mapboard. Mapboards generated from ArcGIS Pro via AIX files cannot be duplicated.

- **Delete:** The Delete button on this toolbar will permanently delete a mapboard from the extension, including mapboards that are synced to an AI file. The Mapboard toolbar's Delete button will not delete an AI or AIX file. However, the link between a synced AI file and Maps for Adobe Creative Cloud will be permanently severed when a mapboard is deleted.

Now with your new mapboard creation knowledge, it is time to move forward to the Compilation panel. This component is where map layers can be added, geo-analyses and map enhancements can be performed, and maps can be downloaded as AI files.

Compilation panel

The Compilation panel is used to add map layers and web maps to a mapboard, as well as edit the map layers' appearance and perform geo-analysis and other enhancements.

The function of the Compilation panel

To enable Compilation panel functionality, a mapboard must be selected from the Mapboards panel. When a mapboard is selected, the Compilation panel shows a preview of the map as it will appear on download. As you add and edit layers on the Compilation panel, the map preview area will update to reflect these changes. Tutorial 2.2 will demonstrate some of the many Compilation panel features using the Minnesota mapboard that you created in tutorial 2.1.

Opening the Compilation panel

The Compilation panel can be opened by clicking the Preview button on the Mapboards panel. The Preview button also appears on the Mapboards panel toolbar after you draw a mapboard. If the Compilation panel is open before you draw or select a mapboard, you will be prompted to return to the Mapboards panel to either create or select a mapboard.

Following are the panel's elements, listed in figure 2.13 as A–P.

A. Add Content button

Clicking the Add Content button (figure 2.14) provides options for adding map data and layers to a map from the Compilation panel. This section describes how each of these options can be used.

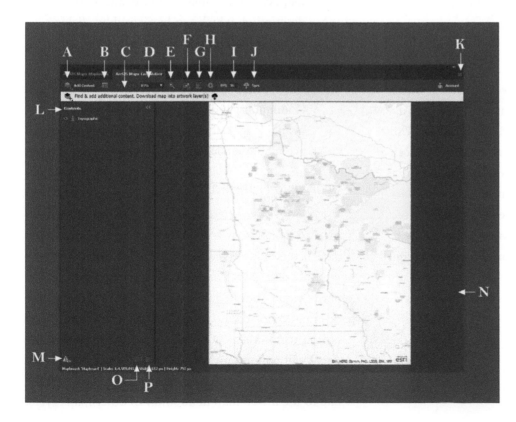

Figure 2.13. The elements of the Compilation panel.

Figure 2.14. Options available for adding content from the Compilation panel.

> **Reminder**
>
> To discover what layers are accessible with each Maps for Adobe Creative Cloud license type, see the ArcGIS Maps for Adobe Creative Cloud Functionality Matrix in table 1.1 in chapter 1 or visit links.esri.com/FunctionalityMatrix.

Add Layers

Add Layers opens the Add Layers window, where users can browse and add data layers hosted in ArcGIS Online. If you are accessing Maps for Adobe Creative Cloud with an ArcGIS Online license, you can also add layers from your own account or organization. Map layers can be added from the following data categories called libraries. Chapter 4 provides more information about these data libraries. Note that data category libraries are different from Illustrator libraries, which you will be leveraging in later tutorials when you learn about the Processes panel.

- **Maps for Creative Cloud:** This category contains map layers that have been specifically selected or created for use in Maps for Adobe Creative Cloud. The content in this category is published by Esri, Esri business partners, and the Esri Community. Some content is public and free for all users, while other content may require additional licensing or consume credits.

- **Living Atlas of the World:** ArcGIS Living Atlas of the World is a large collection of digital geographic information. By using Living Atlas of the World in Maps for Adobe Creative Cloud, you can access ArcGIS Living Atlas and browse among selected map layers, including satellite imagery, boundaries, or other environmental datasets. Some items in ArcGIS Living Atlas may require an organizational subscription to access and may consume credits. Access information is listed in an item's (web map, data layer, image service, and so on) description, which can be viewed by clicking the item's thumbnail image.

- **ArcGIS Online:** Using this option, you can browse map and image layers from the Esri Community and add them to your map. Items available to you will be determined by your account and license type.

Overwrite From Map

Use the Overwrite From Map option to add a web map to a mapboard. A web map differs from map data layers in a few ways. Web maps can consist of multiple data layers and are typically designed to display spatial stories or collections of related geospatial information or data. Using the Overwrite From Map option to load a new web map on the Compilation panel will remove any web maps and layers that have already been added to your mapboard. If you would like to use a web map in addition to other map layers, simply add the web map first. This will allow you to add subsequent map layers.

To add a web map via the Overwrite From Map option:

① From the list of Add Content options, select Overwrite From Map.

② Choose one of the ArcGIS libraries from the Overwrite From Map window, and type a keyword to further filter the results, or select ArcGIS Online URL and enter the web map URL in the text.

③ Select the web map to be added, and click Add to load the web map.

Add Places

Using Add Places, you can search for a place—or many places—and add it to a map as a point of interest, which will display as a circle-shaped point symbol on your map.

① From the list of Add Content options, select Add Places.

② In the Add Places window, enter a place-name, a specific address, or coordinates to search for matching locations.

When searching for places, you can enter a generic name, a specific name, an address, or the latitude-longitude coordinates for a location that you want to add to the map. Your search may result in several locations to choose from. For example, searching for "coffee shop" when your mapboard covers a large city will likely produce dozens, perhaps even hundreds,

of results. ArcGIS Online and Plus license users can select from any or all of these results to add to a map. Complimentary license access allows up to the first 50 results.

Add Layer From file

You can add layers to your map from a file on your computer. Supported file types are SHP, CSV, TXT, GPX, and KML and KMZ.

MAPMAKER TIP

Things to know when adding local files to a map

- If you are adding a shapefile, make sure that all its associated files are zipped into a single .zip file. If the .zip file is missing any component of the shapefile, such as the SHX or PRJ file, the shapefile may not be added to the map.

- CSV and TXT files support various field configurations for geocoding addresses and places. See the "Local data" section in chapter 4 for more information on how to configure these file types for Maps for Adobe Creative Cloud.

- The number of records, or spatial features, that can be added from a local file is 1,000 for Complimentary users and 4,000 for Plus and ArcGIS Online users. If the record number exceeds this limit, an error alert will appear.

- A KML or KMZ file size is limited to 10 megabytes.

B. Attributes button

With the Attributes button 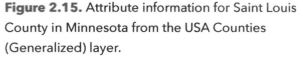 turned on, you can view a map layer's attributes by clicking on the map data in the map preview area.

To view a layer's attributes:

1 Click the Attributes button to enable its functionality.

2 Click on any map feature that has been added to the map.

3 If you click in an area where multiple features overlap, you will be prompted to select which feature's attributes to view. When you select the feature to view or if there are no overlapping layers, the feature's attributes pop-up will appear.

The window displaying the attributes for the USA Counties (Generalized) layer (figure 2.15) also shows an Add button with a plus sign next to each attribute. Clicking this button will add a label to the map that matches that attribute's value for the selected feature. A new layer that contains these additional labels will be created on the Contents panel.

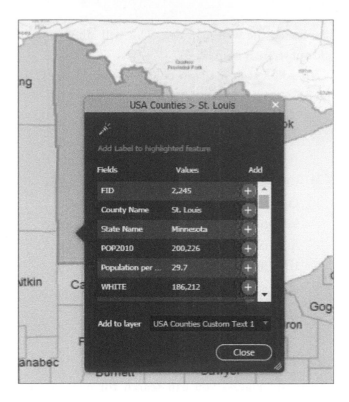

Figure 2.15. Attribute information for Saint Louis County in Minnesota from the USA Counties (Generalized) layer.

C. Undo and Redo buttons

These buttons undo or redo layer-related map actions that are performed on the Compilation panel. The Undo and Redo buttons will not undo or redo pan or zoom actions, nor will the Undo button undo any mapboard adjustments.

D. Quick Zoom

Choose from several percentage zoom levels from the Quick Zoom drop-down list for rapid zooming. Because the Compilation panel displays a map preview, a zoom level different than 100% may impact the map preview's resolution. However, your zoom level on the Compilation panel will not impact the resolution of the downloaded map.

E. Zoom tool

Turn on the Zoom tool to enable the zoom magnifying glass. Click on the map to zoom in one level. Hold the Alt (PC) or Cmd (Mac) key while clicking to zoom out. For more ways to navigate the map preview area, see table 1 on keyboard shortcuts in the introduction to this book.

F. Open Mapboards Panel button

This button will refocus the extension on the Mapboards panel, making it active.

G. Processes button

This button opens the Processes panel. You will learn about the Processes panel functions in chapters 3 and 6.

H. Current Map Settings dialog box

A map's projection can be updated from the Current Map Settings dialog box. In tutorial 4.1, you will reproject a map by updating these settings. For further information on map projections, see the "Map projections" section in chapter 5.

You can update the mapping profile for a selected mapboard on the Current Map Settings dialog box.

The new AIX profile is available only to ArcGIS account types. This setting differs from the Mapboards panel application settings because here you are determining the setting for only a selected unsynced mapboard. Search for "mapping profiles" in the product documentation at links.esri.com /AboutMapsForAdobe.

I. Raster Resolution

Use the Raster Resolution text field to specify the desired pixels per inch (PPI) in your downloaded map. ArcGIS Online, ArcGIS Enterprise, and Plus users can enter a value ranging from 96 to 300 PPI. The PPI can be updated on synced maps as well as for these users. Note that a higher PPI level may result in longer download times. This setting is fixed at 96 PPI for Complimentary users.

J. Sync button

Clicking the Sync button initiates the map download process, resulting in an AI file containing the layers that are added to the Compilation panel. For Plus and ArcGIS Online and Enterprise account users, syncing a map also embeds your mapboard information in the downloaded AI file and into AI files generated from an ArcGIS Pro AIX file. Having this embedded mapboard information allows users to return to the most recent mapboard state in Maps for Adobe Creative Cloud at any time. Also, for these license holders, the Sync button is used to add new map layers to an already synced map, including maps generated from an AIX file. In chapter 3, you will learn more about the advantages of synced maps.

K. Compilation panel Application Settings menu

The Application Settings menu provides access to additional settings and help topics. It also provides the option to share a compiled map to an ArcGIS Online account. You can also access the application-wide Mapping Profile setting, which determines the mapping profile for new mapboards.

L. Contents panel

Layers that have been added to the map will appear on the Contents panel. Users can toggle the visibility and downloadability (syncing) by clicking the respective buttons for each layer. As you work through the tutorials in this book, you will learn how to use the geo-visualization functions available in the layer menu options. The layer menu options are accessed by hovering the pointer over a layer's Options button (ellipsis icon). Not all layer types will have this layer menu, and options will vary depending on the layer and license types.

If you open an AIX file that was created outside Maps for Adobe Creative Cloud, the layers in the AIX export will not appear on the Contents panel or in the map preview area (see "N. Map preview area"). However, any subsequent layers that are added with the extension will appear on both the map preview area and the Contents panel.

M. Non-drawing Layers button

If a layer's scale visibility (zoom level) set by the data author differs from the scale that you specified in the Mapboard settings, it will not be visible in the map preview area and is then considered a non-drawing layer. Non-drawing layers will not be added to a downloaded AI file, and their names will be dimmed on the Contents panel with an alert icon to the left of the name. You can hide these non-drawing layers by toggling off the Non-drawing Layers button.

N. Map preview area

This area previews the map as it will be downloaded. Because it is a preview of a map rather than the dynamic basemap on the Mapboards panel, the map preview on the Compilation panel does not update dynamically as you zoom in and out. Map data will appear less crisp when the map is zoomed to a level that is not 100%. Zooming in or out in the map preview area will not impact the appearance of the artwork in the downloaded AI file.

O. Layer Reorder button

Clicking the Layer Reorder button sorts map layers so that point layers are at the top of the stack, line layers are in the middle, and polygon layers are at the bottom. For more information on points, lines, and polygons as they relate to cartography, see "Cartography corner: Geographic data types" in chapter 3.

P. Delete Layer button

The Delete Layer button deletes selected layers from the Contents panel. To delete multiple layers, press Ctrl (PC) or Cmd (Mac), and click to select multiple layers. Then press the Delete Layer button to remove the selected layers from the map.

Tutorial 2.2: Building a map with the Compilation panel

Scenario

In this tutorial, the second in a two-part exercise, you will be adding layers to the Minnesota mapboard from tutorial 2.1. If you still have your Minnesota mapboard open, you can begin this exercise at step 1. If you have closed Illustrator or deleted the Minnesota mapboard, simply repeat the steps from tutorial 2.1 to draw the Minnesota mapboard again, and then move on to step 1. Since tutorial 2.1 was short, repeating these steps should not take long.

By the end of this exercise, you will have created an AI file with a map of Minnesota counties. Review the learning objectives before you begin.

Learning objectives

- **Doing a quick zoom:** Adjust the map preview area using the Quick Zoom tool
- **Adding map layers to a mapboard:** Add ArcGIS Online content to a mapboard from the Compilation panel's Add Layers window
- **Creating data filters for feature display:** Build custom filters to show a selection of map features
- **Organizing layers:** Manually reorder layers on the Contents panel
- **Removing layers from a mapboard:** Delete map layers from the Contents panel

- **Updating map appearance:** Use the Change Styles layer option to edit map features' appearance
- **Adding map labels:** Label individual layers from the Compilation panel
- **Setting the file location for saving your files:** Update the default directory where your synced files are saved
- **Syncing a map:** Sync and download a map to generate an AI file

Doing a quick zoom

Fit the map to the map preview area

1 Click the Preview button ▦ on the Mapboards panel to open the Compilation panel, and dock the panels.

On the Compilation panel, the basemap that you see in the map preview area probably looks different from the one on the Mapboards panel. This difference is because of the dynamic map rendering on the Mapboards panel versus the static rendering of the Compilation panel. The Mapboards panel basemap is a live view of a web map, which means the map is dynamically drawn at optimal detail depending on the zoom level. On the other hand, the detail of the Compilation panel's basemap was drawn based on the scale you entered and is now static and will no longer update as you zoom in and out. It is a preview of your final map.

2 From the Quick Zoom drop-down list, select Fit On Screen.

The map preview is resized (figure 2.16) so that the entire map extent is visible.

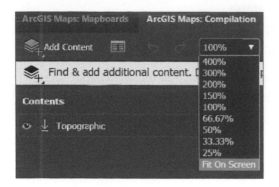

Figure 2.16. Selecting Fit On Screen centers and resizes the map preview to fit the Compilation panel's map preview area.

Adding map layers to a mapboard

Add counties to the map

 On the Compilation panel toolbar, click Add Content, and select Add Layers.

The Add Layers window opens.

 In the Add Layers window, click the arrow to expand the Maps for Creative Cloud library.

 In the Search text field, type **counties**.

This filter will narrow your search to data layers with the text *counties*.

 Click the Filters button, and make sure that Show Vector & Raster Layers is selected.

This selection will search all data types.

 Click the Filters button again to hide it.

 Click in the Search bar to activate it, and press Enter or Return to initiate the search.

 Click the Add button for USA Counties (Generalized) to add the layer (figure 2.17).

Figure 2.17. Searching for data from the Maps for Adobe Creative Cloud library.

8 Close the Add Layers window.

The USA Counties layer is added to the map (figure 2.18).

Figure 2.18. The map preview area with the counties layer added.

Creating data filters for feature display

Add a custom filter to show only Minnesota counties

You have added a data layer that has all the counties for the entire United States. Because your mapboard covers only Minnesota and parts of surrounding states and a little bit of Canada, if you were to download the map now, the resulting map would contain only the counties visible in the mapboard as desired. In this step, you are going to make a custom filter that will allow only the Minnesota counties to display (figure 2.19).

 On the Contents panel, hover the pointer over the layer's Options button to show the USA Counties (Generalized) layer options menu. Then select Filter from the menu options.

 From the first drop-down list, select the State field.

You will need to scroll to nearly the bottom of the list to find this field.

 Select Is from the selection clause drop-down list (middle option).

 Type **Minnesota** in the final text field.

Note that it is case sensitive.

 Once your filter is set, click Apply Filter.

Figure 2.19. This filter is set to display only the counties that have Minnesota as their name in the State field.

Add USA states to the map

To offer spatial context for your map readers, you will now add a layer of the states that surround Minnesota.

 Once again, click Add Content, and select Add Layers.

 Select Maps for Creative Cloud as the library, and type **states** in the Search text field. Then press Enter or Return.

3 Scroll until you see USA States (Generalized), and click that item's Add button to add the layer.

4 Close the Add Layers window.

Organizing layers

Reorder layers manually

The most recently added layer, USA States (Generalized), is at the top of the layer stack on the Contents panel. You can also see that it sits above the Minnesota counties in the map preview area. To prevent the states from obscuring the counties, you will reorder the layers manually.

1 On the Contents panel, drag the USA States (Generalized) layer name below the USA Counties (Generalized) layer so that a blue line appears beneath it (figure 2.20), and then release to place the states below the counties.

Figure 2.20. The blue line indicates where the USA States (Generalized) layer will be placed when dragging it below the counties layer. After the mouse button is released, the layer is placed in that order.

Add Canadian provinces to the map

For even more spatial context for your map readers, add the provinces of Canada to the map.

1 Just as you have been doing in previous steps, open the Add Layers window. Then select the Living Atlas of the World library. To further narrow your search, select the Boundaries group within this library.

2 Type **countries** in the Search bar, and press Enter or Return.

3 Scroll until you find the World Countries (Generalized) layer (figure 2.21), add this layer to your map, and close the panel.

By adding this polygon layer of world countries, your final map will have symbolization for Canada.

Figure 2.21. Searching for countries from the ArcGIS Living Atlas Boundaries group.

4 Drag the layer just above the Topographic layer on the Contents panel.

Checkpoint

At this point in tutorial 2.2, there should be four layers on the Contents panel in the order shown: counties, states, world countries, and topographic. Although the color of your polygons may differ, the Compilation panel should appear as it does in figure 2.22, with all four layers added.

Figure 2.22. The Compilation panel is now complete.

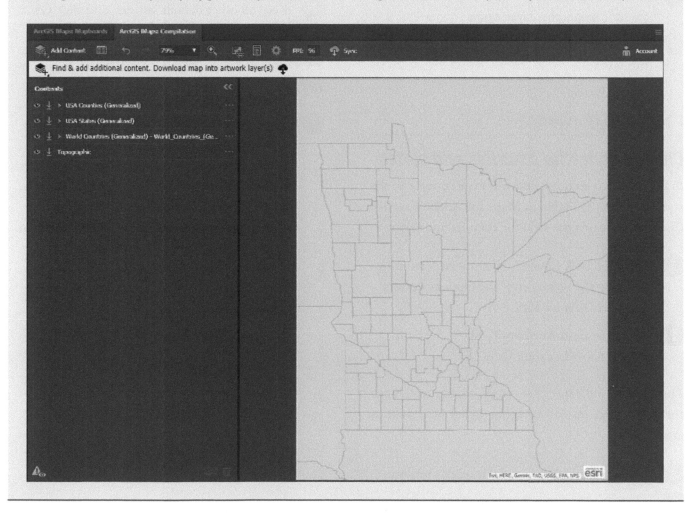

Remove the raster basemap

The raster basemap is now covered completely with the map layers that you added in this tutorial. The basemap is no longer useful for this map. It is time to remove it.

 Hover the pointer over the Options button next to the Topographic layer name.

2 Select Delete from the layer options to remove this layer.

Updating map appearance

Change styles for counties and states

In the next two sections, you will explore some of the styling options available in Maps for Adobe Creative Cloud by changing the style for counties and states.

1 Open the USA Counties layer menu options by hovering the pointer over the Options button, and select Change Style.

2 For Choose An Attribute To Show, keep the default Show Location Only.

This option indicates that every county will have the same style, instead of being styled by a field or attribute.

3 Under Select A Drawing Style, click Options.

4 Click Change Symbol Style to open the color picker.

You will have the option to change the fill and outline colors by clicking their respective tabs in the color picker.

5 With the Fill tab selected, enter **#E9C8FC** as the hexadecimal color value.

6 Click the Outline tab, and type **#AC99AD** as its hexadecimal color value.

7 Keep all other default settings, and click OK.

Change styles for the remaining layers

1 Change the fill and outline colors for the USA States (Generalized) and World Countries (Generalized) layers following the same process, but this time set the fill color for each of these two layers to **#EBEBEB** (figure 2.23), and set the outline color for these layers to **#D6D6D6**.

Figure 2.23. The map preview area has all counties set to the same color.

Add the country boundary line

The USA-Canada boundary line is the final layer that you will add to the Minnesota mapboard.

 Once again, open the Add Layers window, and select ArcGIS Online as the library to search.

 In the Search text field, type **Tutorial_2_2_ International_Boundary**, and press Enter or Return.

 If you do not see the item thumbnail (figure 2.24), uncheck all the boxes in the Add Layers window's search filters.

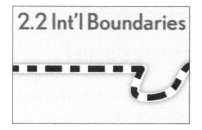

Figure 2.24. The thumbnail for the international boundaries layer you are adding.

 Click the item's Add button to add it to the Minnesota mapboard, and close the Add Layers window.

Add labels to features

 On the USA Counties (Generalized) layer, hover the pointer over the Options button to expose the layer options, and click Manage Labels.

The Label Options window opens.

 In the Label Options window, check the Show Overlapping Labels box so that all county names appear on your map.

 Change the text size from 13 to 8 by typing **8** or selecting 8 in the corresponding text field.

 Switch the label fonts from bold to regular by clicking the Bold button (B) next to the text size.

At this point, the label option settings should appear as shown in figure 2.25, with the name in 8 pt. regular and showing overlapping labels.

Figure 2.25. Updating the label to regular size 8 and showing all overlapping labels.

 Click OK to finish adding the county labels.

Setting the file location for saving your files

Set the default location

1 On the top right of the Compilation panel, click the Menu button (three-lines icon) to examine the settings.

2 Click Settings.

3 With the General tab highlighted, click the Output Directory Options button.

4 Browse and create folders to set the default location to your computer in a location that you will remember, such as EsriPress \MappingByDesign.

5 Click OK to close the Settings window.

Syncing a map

Sync to download your map

Now you are ready to download your Minnesota counties map as an AI file. When you download the map, the AI file that is created will also be synced to the Maps for Adobe Creative Cloud extension. Before syncing, make sure you note the location where the AI file will be saved, as in EsriPress\MappingByDesign, so that you can quickly locate it for tutorial 3.1. You can specify where you want your Maps for Adobe Creative Cloud files to be saved on the Mapboards panel application settings and on the Compilation panel application settings General tab. In tutorial 3.1, you will explore the AI file you created from this sync.

 Click Sync on the Compilation toolbar to download your map. Click OK to confirm.

Note that you may see your map appear in Illustrator before it is completely processed by the extension. It is important to let this process finish before interacting with the AI file. Depending on your computer's processing speed, the download can take a few moments, so be patient while your map is created.

Once the map is synced, you will see a message on the Action bar letting you know it is ready for design. After the file is ready, you are free to explore your Minnesota.ai map that you have just downloaded, but do not make any changes to it. It will need to remain in the current state for the next tutorial.

A tour of the Processes panel

This chapter concludes with a tour of the third and final panel, the Processes panel, which contains several map enhancement functions and tools. You will begin using the Processes panel to enhance your maps in tutorial 3.2 by automatically generating a map legend in your synced Illustrator file.

1 Click the Processes button on the Compilation panel, and dock the panel.

Some of these processes can be run during a map sync, whereas others are post-processing tools that you can use on an AI file only after you have downloaded it through Maps for Adobe Creative Cloud (see chapter 7 for running processes tools on AIX-generated files).

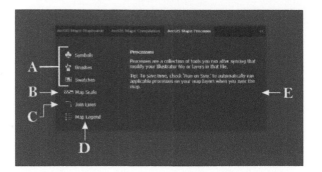

Figure 2.26. The Maps for Adobe Processes panel.

Following are the panel's elements, listed in figure 2.26 as A–E.

A. Custom symbols, brushes, and swatches processes

The first three items on the Processes panel are the symbols, brushes, and swatches replacement processes. These processes can be used to automatically replace default map symbology with styles from an Illustrator Symbols, Brushes, or Swatches library. You may set up these symbol replacement processes to run either during a map sync or on an already-synced AI file, including those generated from ArcGIS Pro AIX file exports. You will be using these symbol replacement options during the tutorials in chapters 4 and 6. Arc-GIS Pro users can follow along in chapter 7 to learn about running these processes from maps shared as AIX files.

B. Map scale process

You can add a scale bar on AI files created with Maps for Adobe Creative Cloud using the map scale process.

C. Join lines process

You can concatenate unconnected line segments on synced AI files using the join lines process. The operation joins selected paths that are in the same layer at their touching endpoints within a distance tolerance of 0.005 points.

D. Map legend process

You can create a legend in a synced AI file using the map legend process. To build a map legend in Illustrator, turn on the visibility of the layers to be included in the legend, and turn off the remaining layers' visibility. Then click Create Legend to run the Map Legend tool.

E. Process description area

You will find each process's description in this area. Click the process's tab to the left to show the description.

2 Close Illustrator.

CHAPTER 3
THE ADVANTAGES OF SYNCED MAPS

WHAT IS THE DIFFERENCE BETWEEN SYNCING and downloading a map?

Sync and *download* are terms that are often used interchangeably when discussing the Maps for Adobe Creative Cloud workflow. When you are signed in to Maps for Adobe Creative Cloud with a Plus license or ArcGIS Online or ArcGIS Enterprise, an AI file that is downloaded—or created—from the Compilation panel will also be synced with Maps for Adobe Creative Cloud, which creates a beneficial link between the AI file and Maps for Adobe Creative Cloud. As you will learn in chapter 7, an AIX file generated in ArcGIS Pro will also be converted to a synced AI file when you open it in Illustrator. In either workflow, ArcGIS Pro–to–Illustrator or the extension-direct workflow, the link between a synced AI file and Maps for Adobe Creative Cloud connects the AI file to the most recent state of the synced mapboard. This connection benefits mapmakers in several ways.

After reading through this chapter's first tutorial, tutorial 3.1, you will have firsthand experience with some of the advantages of AI files synced to Maps for Adobe Creative Cloud. In tutorial 3.2, you will leverage even more of these advantages by using Maps for

Adobe Creative Cloud to add new layers to a synced mapboard. Both tutorials in this chapter will build on the Minnesota.ai map that you created in chapter 2.

CARTOGRAPHY CORNER

Geographic data types

As you continue reading through this book and working through the tutorials, you will encounter references to point, line, and polygon data. These three terms refer to the different GIS vector data types. When it comes to map design, understanding the differences and advantages of point, line, and polygon data—or map layers—will greatly assist map designers, even those who do not consider themselves GIS experts. In graphics editing programs such as Adobe Illustrator, a map's points, lines, and polygons are essentially the building blocks, or paths, for the aesthetic design.

Point data

Points are an excellent way to display a geographic feature that is too small to depict as a line or polygon. A single GIS point location consists of one pair of x,y coordinates (x,y,z coordinates if the map is 3D). In a GIS, the x typically indicates the longitude, and the y indicates the latitude. When z is included, it indicates the position's elevation.

In map design, it is common for cartographic point symbols to indicate category or hierarchy. For example, in a small-scale (zoomed out) map that features cities and towns, a large city point symbol will be more prominent than the point symbol for a small town.

Contest winner Noah Walker's map of Arkansas (figure 3.1) uses points to represent selected cities. The capital city of Little Rock is symbolized by a star point symbol, whereas the remaining cities are represented by smaller circular point symbols. This strategy indicates the hierarchical difference between two city categories. Walker's Arkansas map was created using Maps for Adobe Creative Cloud.

Points on maps can also represent more than one object. Clustering is a cartographic technique that generalizes many points of interest into a larger symbol indicating a cluster of geographic points. Often the number of points represented in a cluster is indicated by the point's label. On the *Rock Climbing & the Solar Eclipse* map (figure 3.2), notice that the sizes of the points differ, indicating the general amount of first ascents each point represents. This technique is called graduated symbology, and you will be using it in tutorial 4.1.

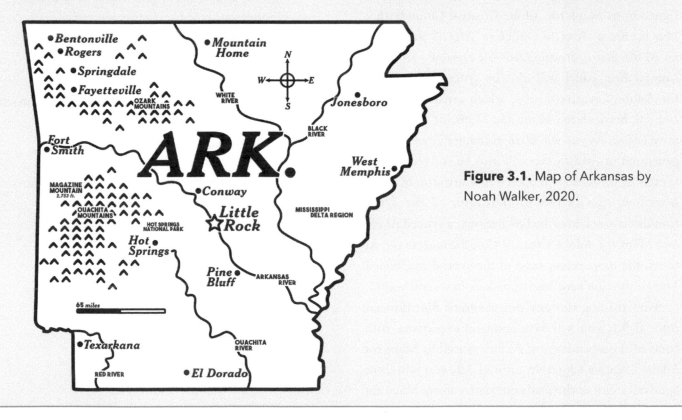

Figure 3.1. Map of Arkansas by Noah Walker, 2020.

Mapping by Design: A Guide to ArcGIS Maps for Adobe Creative Cloud

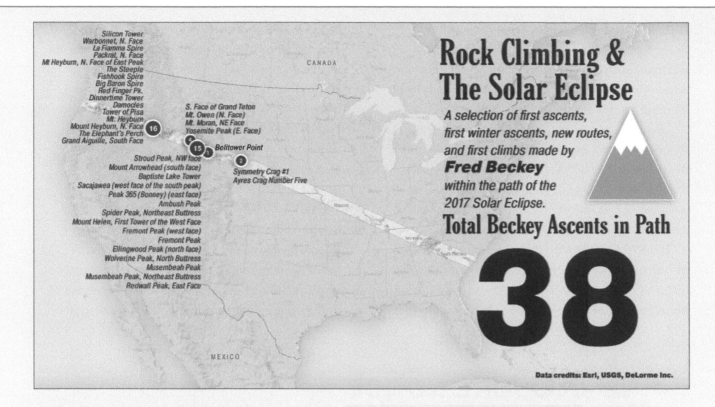

Another method for using a single point symbol to represent multiple geographic phenomena is with a dot density map. For example, a map that uses one dot to represent a certain number of people over a particular area is called a dot density map. On dot density maps, dots are scattered across their respective regions for the purpose of illustrating the density of the mapped geographic phenomenon. A nice addition to the dot density technique is to ensure that dots are closer to geographic features where they are more likely to occur. For example, you can map dots that represent people close to towns and cities rather than homogenously scattering the dots across a polygon.

Figure 3.2. I created this map using point clustering to show selected climbing ascents by Fred Beckey that occurred within the path of the 2017 solar eclipse across North America.

Line data

In cartography, lines represent features that are conceptually long and narrow, and too small to be depicted as areas. In GIS datasets, streets and linear water features, such as creeks and rivers, are commonly represented by line data. These data are made up of nodes that are joined by lines into a single feature. A river, especially when mapped at the large scale, can be represented by a polygon.

However, linear data are optimal for mapping linear features at a small scale. Polygonal rivers will lose much of their detail at the small scale and add unnecessary complexity to the map as well as increase the file size.

Lines are also used to represent map features that do not have an aerial shape but have length. Boundary lines, such as the USA-Canada border that you added to your map in tutorial 2.2, and isolines such as elevation contours are excellent examples of this kind of linear data.

The *I-10 Closure Map* by California Department of Transportation senior graphic designer and contestant Andrew Pham (figure 3.3) was created using Maps for Adobe Creative Cloud. This winning map shows the temporary closure of a small section of one of California's major interstates, as well as the official proposed detour during this closure period. Pham cleverly symbolized the closure segment with a thick, solid red line underneath an attention-grabbing brighter dashed line. The detour segment is green, telling readers that this is the best route during the closure period.

Polygon data

Polygon data represent the aerial coverage of geographic features. Like line data, polygon features are made up of nodes connected into a single feature. However, unlike lines, the nodes of polygons are connected into a closed polygonal

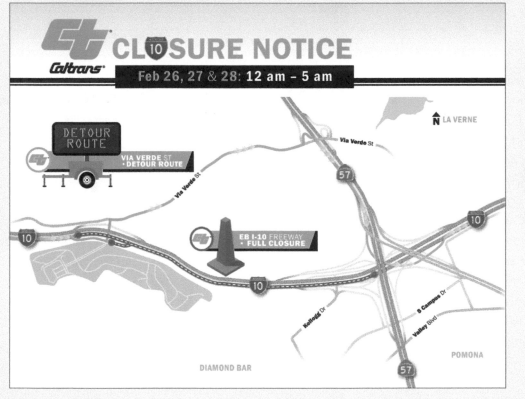

Figure 3.3. I-10 closure map by Andrew Pham, 2020.

shape. The following map by contest winner Katelin Volkanovski uses bright-yellow polygons to show the habitat distribution of the yellow-tailed black cockatoo, a variety of Australia's black cockatoo. In addition to the bird habitat polygons, the country of Australia also comes from polygon data. Volkanovski created this map (figure 3.4) using Maps for Adobe Creative Cloud.

Figure 3.4. *Where Might I See the Yellow-Tailed Black Cockatoo?* by Katelin Volkanovski, 2020. The map shows the distribution of one of six varieties of Australia's black cockatoo.

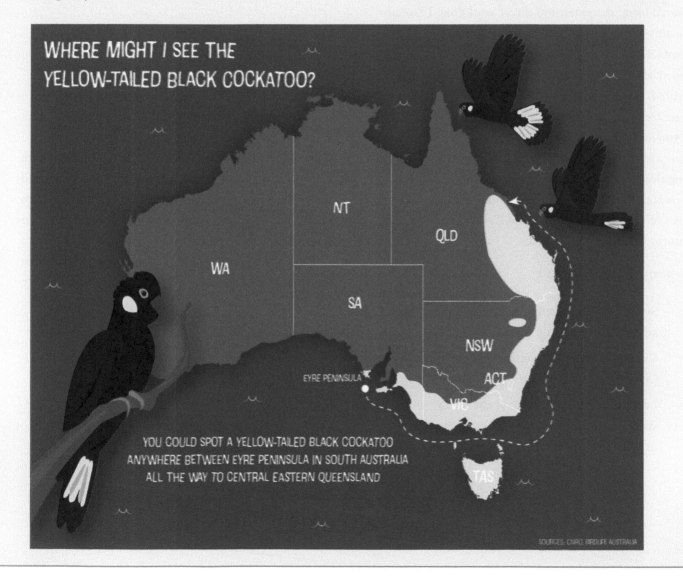

WHERE MIGHT I SEE THE
YELLOW-TAILED BLACK COCKATOO?

NT

QLD

WA

SA

NSW

ACT

EYRE PENINSULA

VIC

YOU COULD SPOT A YELLOW-TAILED BLACK COCKATOO
ANYWHERE BETWEEN EYRE PENINSULA IN SOUTH AUSTRALIA
ALL THE WAY TO CENTRAL EASTERN QUEENSLAND

TAS

SOURCES: CSIRO, BIRDLIFE AUSTRALIA

Tutorial 3.1: Exploring a synced map file

Scenario

Tutorial 3.1 is an exploratory exercise investigating the Maps for Adobe Creative Cloud settings on a synced mapboard and the layer structure of an AI file created using the extension-direct workflow. Throughout this exercise, you will be using your synced Minnesota.ai file that you created in tutorials 2.1 and 2.2 and synced in tutorial 2.2. You will learn how AI files created using Maps for Adobe Creative Cloud are organized as you examine the Minnesota.ai file layers. If you still have Illustrator open, it is recommended that you close and reopen the program before beginning this tutorial.

Note: Because this tutorial highlights the advantages of synced files, a Plus, ArcGIS Online, or ArcGIS Enterprise license is required.

Learning objectives

- **Synced mapboards:** Become familiar with the advantages of synced mapboards

- **Contents Panel layer settings:** Learn about the Compilation panel's layer sync setting

- **Layer organization:** Understand the layer structure of an Illustrator file created directly in Maps for Adobe Creative Cloud

Synced mapboards

Open the synced map

1 Open Illustrator, and then sign in to Maps for Adobe Creative Cloud (Window > Extensions > ArcGIS Maps for Adobe).

Do not open the Minnesota.ai file yet.

2 While keeping the Mapboards panel in view, open your Minnesota.ai map (File > Open). Browse to EsriPress/MappingByDesign or the location on your computer where you saved your folder, and open the file Minnesota.ai.

When opening your file, the extension's window might automatically minimize or become placed behind Illustrator's user interface. If this occurs, reselect Maps for Adobe Creative Cloud from the Illustrator Extensions menu to pull the extension into view again.

Once the Minnesota.ai file is open, the Mapboards panel displays the Minnesota mapboard extent (figure 3.5).

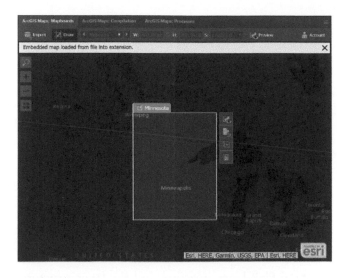

Figure 3.5. When the Minnesota.ai file is open, the synced mapboard extent will appear on the Maps for Adobe Creative Cloud Mapboards panel. Your mapboard may be different depending on the basemap you have selected, which is fine.

3 Open the Mapboard Options dialog box by clicking the Mapboard Options button on the vertical toolbar to the right of your mapboard. The Mapboard Options button is the second button from the top.

Notice that the fields on the dialog box are dimmed. The reason that these fields are no longer editable is that settings such as size, scale, and extent are permanently synced to this mapboard and can no longer be changed.

4 Click OK to close the dialog box.

Explore syncing and visibility

Just like the layer visibility setting in Illustrator, the visibility of the data layers on the Compilation panel's Contents panel can be toggled on or off. When a layer's visibility is toggled off, the visibility icon (eye) disappears. Maps for Adobe Creative Cloud also has a sync setting for each layer on the Contents panel. This sync setting, shown by a down arrow, specifies whether a layer will be added to a map when it is synced.

When you opened the synced Minnesota.ai file, the Compilation panel's Contents panel became populated with all the layers that were present on the Contents panel during the AI file's most recent sync.

1 Open the Compilation panel, and turn on the sync setting for the counties layer by clicking the blank rectangle directly to the right of the visibility icon for the USA Counties (Generalized) layer.

A down arrow will appear (figure 3.6).

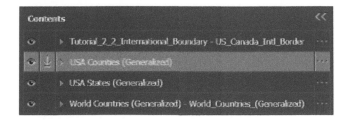

Figure 3.6. The sync setting is on for the USA Counties (Generalized) layer.

Maps for Adobe Creative Cloud automatically turns off a layer's syncing after a mapboard is synced to prevent unwanted duplicate layers in subsequent syncs. A layer's sync setting can be manually toggled on at any time. For this tutorial, keep the sync setting of the counties layer on and the remaining already synced layers off.

MAPMAKER TIP

Syncing with visibility turned off

On the Compilation panel's Contents panel, a layer's visibility and sync setting are somewhat linked. If you turn off layer visibility, the sync setting will also be turned off. But there will be times when you want to download layers with their visibility turned off in your resulting AI file. You can do this, too. If you want to sync a layer while keeping its visibility toggled off, first turn the visibility off and then turn the sync setting on.

Layer organization

Compare the layer structure

An AI file that is created using Maps for Adobe Creative Cloud will be organized with a ready-to-design layer structure that corresponds to the layer arrangement set up on the Compilation panel's Contents panel. However, the layer structure between the Contents panel and the downloaded AI file may differ slightly depending on how the data were organized in the original map data. In this section, you will compare the similarities and differences between the Minnesota.ai file's layer structure and the layer structure from the Compilation panel's Contents panel.

1   In Illustrator, open the Layers panel, and expand the Minnesota_Elements layer by clicking the arrow to the left of the layer name.

This is a parent layer containing a sublayer for a vector Esri logo and a sublayer for any map credits that were included by the original data's author.

2   Expand the Minnesota_Sync_1 parent layer to reveal its sublayers.

Whenever you sync a map, a parent layer for that sync is created with the naming convention Mapboard-Name_Sync_1, in which the mapboard name is at the beginning and the chronological sync number is at the end. For example, your parent layer is called Minnesota_Sync_1 because the name of the synced mapboard is Minnesota, and this parent layer contains sublayers added during the first sync.

Category sublayers

As you will learn in tutorial 3.2, map data layers can contain categories, which are useful for cartographic design and organization. For example, on the Compilation panel, you can symbolize a transportation layer by its categories. For example, the transportation categories could be railroads, ferry routes, city streets, and highways. When using the Maps for Adobe Creative Cloud Change Style feature layer options menu to symbolize the transportation layer by these categories, Maps for Adobe Creative Cloud will automatically create sublayers for the synced AI file, one sublayer for each category. These sublayers will be in a parent layer called Transportation. This automatic layer organization makes it simple to edit and design layers in Illustrator.

Checkpoint

You have just completed tutorial 3.1. You have learned about the basic behavior of synced mapboards for maps created using Maps for Adobe Creative Cloud. Your Contents panel on the Compilation panel should match figure 3.6, with USA Counties synced, and your Illustrator Layers panel should match figure 3.7, with all the synced layer's sublayers, although the colors may differ.

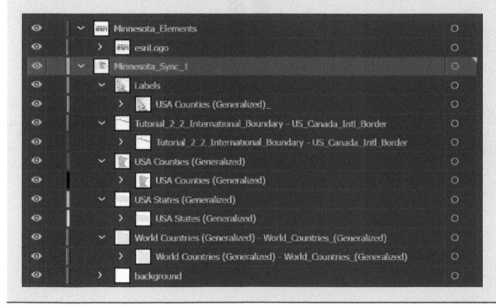

Figure 3.7. After syncing, the Illustrator layers panel will have the same layers as shown in the figure.

Making changes to synced mapboards

- **Add and remove layers, labels, and geospatial analyses:** Although changes cannot be made to a synced mapboard's settings (size, scale, spatial extent, and name), you can add and remove layers and resync layers to a synced AI file. You can also add labels and perform geospatial analyses. A new sync will add these updates to the AI file as new sublayers in a new synced parent layer.

- **Modify the Contents panel:** If you remove a layer from the Maps for Adobe Creative Cloud Contents panel, it will not remove the layer from an AI file.

- **Duplicate or delete mapboards:** On the Mapboards panel, you can duplicate or delete synced mapboards. By duplicating a mapboard, you are creating an unsynced mapboard, which copies all layers that still exist in the mapboard that is being duplicated. In a duplicated mapboard, you can make changes to the size, scale, and spatial extent if the duplicated mapboard is not yet synced. Be aware that deleting a synced mapboard from the Mapboards panel will permanently remove the link between your downloaded AI file and Maps for Adobe Creative Cloud. Deleting a synced mapboard will not delete an AI file.

- **Review:** When you export an AIX file from ArcGIS Pro or ArcGIS Map Viewer, the mapboard associated with the AIX file is automatically synced.

Data classification

Natural breaks (Jenks)

Developed by cartographer George Frederick Jenks, this algorithmic data classification groups data into categories that have the smallest amount of statistical variance (figure 3.8). Because of this minimal variation within classes, the natural breaks classification is suitable for grouping features that are similar for a given numerical attribute. This method is especially useful for a dataset with high variance.

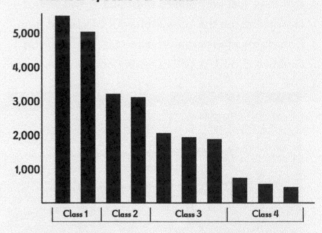

Figure 3.8. Each bar in the chart represents a theater. The classes represent natural breaks, where bars are designated for classes with the least amount of variance in the number of seats per theater.

Equal interval

Equal-interval data classification (figure 3.9) divides the mapped attribute's values into equally sized subranges. First, the entire range of the data attribute is identified. For example, if you are mapping the seating capacity of a city's 10 theaters, where the smallest theater can hold 500 people and the largest can hold 5,500, the range of seating capacity is 5,000 because 5,500 − 500 = 5,000. To group the seating capacity into five subranges of equal size, divide the full dataset's range of this attribute by the number of desired groups, 5,000 ÷ 5 = 1,000. With each subrange as 1,000, the five classes are 500 to 1,500; 1,501 to 2,500; 2,501 to 3,500; 3,501 to 4,500; and 4,501 to 5,500.

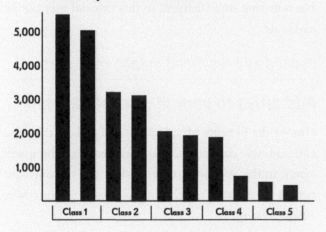

Number of Seats Per Theater

Figure 3.9. The classes in this bar chart represent equal intervals. This figure also represents classes divided into quantiles.

Standard deviation

Standard deviation class breaks are equal-value ranges determined by the classes' standard deviation from the attribute's mean value. Standard deviation is valuable when the mean of the mapped attribute is significant to the map's story. One example might be a choropleth map showing an area's income by district polygons in which mean income is crucial to the map's purpose.

Quantile

Classes grouped by quantile contain an equal number of features. For a map with 50 counties, each class would have 10 counties when using a quantile classification with five classes. Quantile is useful for mapping linearly distributed data. Figure 3.9 can also be considered a quantile representation of data; each class represents quantiles in increments of 20. For example, class 1 contains all theaters in the top 80th percentile for number of seats. Or only 20 percent of theaters are in this class.

Manual breaks

To define your own classes, use manual breaks. Often, mapmakers will use natural breaks or standard deviation classification, and then slightly adjust the classes by rounding them, which is another form of manual breaks. You will be manually adjusting natural breaks classes in tutorial 3.2.

Tutorial 3.2: Adding customized layers to a synced mapboard

Scenario

In this tutorial, you will add a new data layer to your synced Minnesota mapboard. You will also apply a custom filter to this new layer so that only the features necessary for this map will be added to your existing Minnesota.ai file as you perform a second sync for the Minnesota mapboard. In addition to new layers, you will add a duplicate of the counties layer. However, this time you will symbolize the counties layer by quantitative categories. Finally, you will use the Processes panel to add a map legend to your Minnesota.ai file.

Learning objectives

- **Synced and resynced mapboards:** Add new and duplicate map layers to a synced mapboard

- **Data filter review:** Create custom filters for map layers

- **Data-driven mapping:** Style layers with custom symbology using data-driven methods

- **Map legend process:** Add a map legend to an AI file using the map legend process

To begin this tutorial, make sure you have Illustrator open and are signed into Maps for Adobe Creative Cloud.

Note: It is recommended that you use your Plus or ArcGIS Online (GIS Professional or Creator) license or the free trial license that you received with this book. You may continue with a different license type but note that some features in this tutorial may not be available.

Synced and resynced mapboards, part 1

Add cities to your Minnesota.ai map

One of the benefits of synced mapboards is that you can add new data to them by performing subsequent syncs. In this section, you will add cities to your Minnesota mapboard, which you later sync to your Minnesota.ai map.

1. If you do not already have it open, open your Minnesota.ai file, and with the file selected as the active file tab in Illustrator, open the Maps for Adobe Creative Cloud Compilation panel.

From the Compilation panel toolbar, click Add Content, and select Add Layers.

3 In the Add Layers window, choose the Maps for Creative Cloud library (figure 3.10).

4 In the Search input field, type **Major Cities**, and press Enter or Return.

5 Click the Add button for the USA Major Cities item to add this point layer to your map, and then close the Add Layers window.

Figure 3.10. Searching for "Major Cities" from the Maps for Creative Cloud library in the Add Layers window.

Duplicate synced layers from the Compilation panel

In this section, you will set the USA Major Counties layer to be resynced.

1 Confirm that the down arrow appears next to the USA Counties (Generalized) layer.

This arrow indicates that the layer will be added to the AI file during the next sync. The only layers on the Contents panel that should have sync turned on (figure 3.11) are the USA Major Cities layer that you added in the first section of this tutorial and the USA Counties (Generalized) layer.

Figure 3.11. The Contents panel contains the layers added in tutorial 2.2 and the new major cities layer added in the current tutorial. Syncing is on for the major cities layer and the counties layer.

Create a custom filter to show only large cities in Minnesota

In this step, you will create a custom filter for the USA Major Cities layer to show only the cities in Minnesota with a population of 85,000 or higher.

 Expose the USA Major Cities layer options by hovering the pointer over that layer's Options button. Select Filter from the menu options.

You will create a new filter to show only the Minnesota cities from this layer.

2 On the filter dialog box, select State from the first drop-down list, and keep the second drop-down set to is. Then type **MN** as the value (case sensitive), which is the abbreviation used in this layer for Minnesota. Do not apply this filter yet.

3 On the filter dialog box, click Add Another Expression.

4 Select Population (2015) from the leftmost drop-down list.

This field contains the US Census's official population estimates.

5 In the second drop-down list, choose Is At Least.

 In the value field, type **85000**.

This process will create a new filter showing only cities with a population of at least 85,000 people for the year 2015 (figure 3.12).

Once you apply this filter, the map will display all the city points in the USA Major Cities layer that have MN as their State attribute and where the Population (2015) attribute contains a value of at least 85,000. In other words, when you sync this layer to your map with this filter applied, only the cities in Minnesota with a population of 85,000 people or more will be added to your Minnesota.ai file.

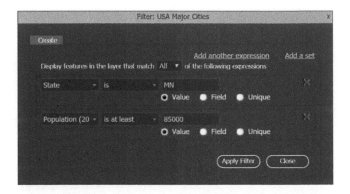

Figure 3.12. This filter will display the major cities in the USA Major Cities layer that are in the state of Minnesota and that have a population of 85,000 or more.

 Click the Apply Filter button to apply the filter.

Now you should see new points on your map in the map preview area. The points may vary in color. In fact, if you click the drop-down arrow to the left of the USA Major Cities layer on the Contents panel, you will see the root of this color variation; the person who published this data to ArcGIS Online symbolized the points by their population class, which is an attribute in this dataset indicating city population. In the next step, you will update this symbolization so that all points are identical in appearance.

Data-driven mapping

Modify styles

In this section, you will update the appearance of the city points so that each point is styled the same.

1. Expose the USA Major Cities layer options again by hovering the pointer over that layer's Options button, and select Change Style.

2. In the Choose An Attribute To Show drop-down list, select Show Location Only.

This selection applies the same symbology to all the points in the layer.

3. Under Select A Drawing Style, click Options.

4. Click Change Symbol Style to open the style window. Then click the Fill tab, and type the hexadecimal value **#919191** in the property's text area to update the city point color to dark gray (figure 3.13). Click OK to close the Style window.

5. Click OK on the Change Style panel to apply the style and return to the Contents panel.

Now the style will be applied.

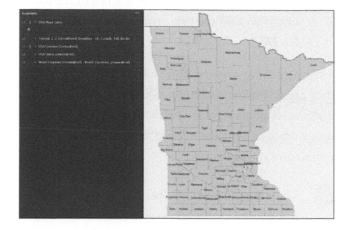

Figure 3.13. Updating the fill color for the points layer results in a map with all the cities in the same color.

Create a choropleth layer

For this section, you are going to symbolize the counties by population density by creating a choropleth map. Choropleth maps help visualize a quantitative characteristic across (usually contiguous) geographic areas represented by polygons. Read more about choropleth maps in "Cartography corner: Choropleth maps" at the end of this chapter.

1. From the USA Counties (Generalized) layer options, select Change Style.

The default symbology for this layer is set to Show Location Only, which means that all the counties are symbolized by the same color.

In the Choose An Attribute To Show drop-down list, scroll toward the bottom of the list until you find the two instances of Pop. Per Sq. Mi.

Although these two instances have the same name, they are distinct and contain different values. The first instance is generated from US Census 2017 population numbers, and the second option contains 2010 population estimates.

 Select the first instance because it is the most recent information from this dataset.

By choosing this attribute, your counties will automatically be styled as a choropleth map. Make sure that you are not selecting the Population (2017) attribute, but rather the first instance of Pop. Per Sq. Mi.

④ Click Options in the Counts And Amounts (Color) section to customize the county category classification. The classification method that is automatically applied is natural breaks. Keep this setting as the classification for your counties layer.

See "Cartography corner: Data classification" in this chapter to learn more about these classification options.

⑤ Manually adjust the two largest categories to give them round numbers.

 Click directly on the top category's value (787) next to the color ramp to make it editable.

⑦ Type **800**, and press Enter or Return.

Now all counties with a population density higher than 800 people per square mile will all be symbolized by the same color.

⑧ Click the next-highest value (395), and update it to **400**.

Now the second category has all counties with a population density ranging from greater than 400 to 800 people per square mile (figure 3.14).

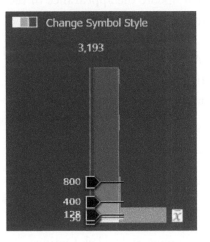

Figure 3.14. The updated classes for the USA Counties choropleth map.

⑨ Click OK at the bottom of the panel to save your changes.

Checkpoint

You have a few more series of steps before you perform a second sync on the Minnesota mapboard. Let's check in before the final sync. You should have a choropleth (color-coded) USA Counties (Generalized) layer that looks similar to figure 3.15. At this point, the counties are still labeled, but you will be removing the labels from the map preview before you complete this tutorial.

There should also be a point layer showing the largest cities in Minnesota. This USA Major Cities point layer and the choropleth counties layer should be the only layers on the Contents panel that are set to be synced.

Figure 3.15. After symbolization, the Contents panel and map preview area should appear similar to the figure.

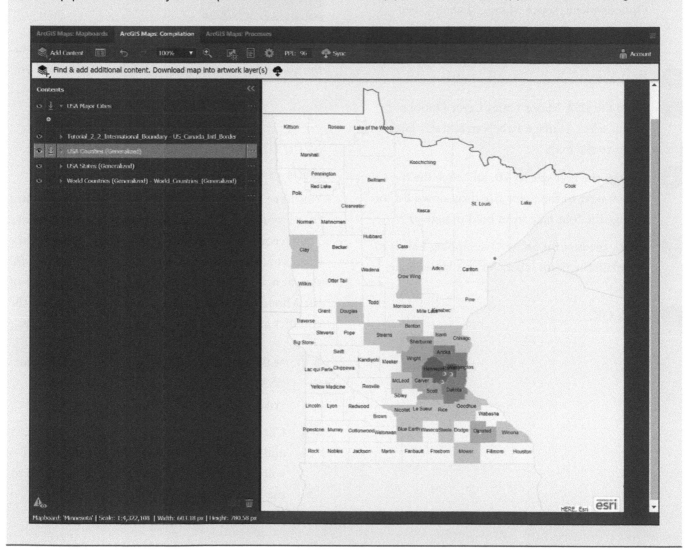

Update layer labels

Because your first sync of this Minnesota mapboard contained county labels, you will not need to add county labels again. In this step, you will turn off labels for the USA Counties (Generalized) choropleth layer, and then you will add labels for the new USA Major Cities layer.

① From the USA Counties (Generalized) Layer Options menu, select Manage Labels. Then turn off the labels by unchecking the box next to Label Features.

② Click OK.

③ From the USA Major Cities Layer Options menu, select Manage Labels to add this layer's labels to the map.

④ Change the text size to **10**, and click the *B* button next to the point size drop-down list to change the font face from bold to regular.

⑤ Check the box for Show Overlapping Labels to reveal all the point labels for this layer (figure 3.16).

⑥ Click OK.

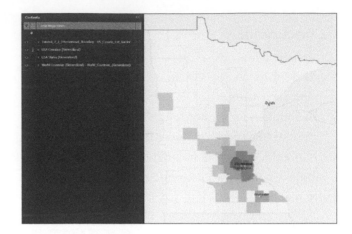

Figure 3.16. County labels have been turned off for the second sync, and city point labels are turned on.

Synced and resynced mapboards, part 2

Sync the Minnesota mapboard again with your new updates

You have now prepared your Minnesota mapboard to be synced for a second time with new updates. Your second sync will add a new parent layer for the sync, containing the city points layer and a layer for this new layer's labels, as well as a new instance of the Minnesota counties layer. In this new instance, the counties are symbolized by their population density.

① Make sure that only the USA Major Cities and USA States (Generalized) layers are set to be synced.

② Click Sync on the Compilation panel toolbar, and wait for the map to finish syncing, which will also result in the new layers being added to the Minnesota.ai file.

Add a legend using the Processes panel

When you performed the second sync in "Synced and resynced mapboards: Multiple syncs," a parent layer for this sync called Minnesota_Sync_2 was added to your Minnesota.ai file.

1 If this parent layer is not expanded already, expand it to reveal its sublayers that contain

the new layers: Labels, USA Major Cities, and USA Counties (Generalized) (figure 3.17).

In the USA Counties (Generalized) layer (of the Minnesota_Sync_2 parent layer), notice that there is a sublayer for each of the value ranges that you set up when you designed the choropleth map in Maps for Adobe Creative Cloud. Because you want your map readers to easily decipher the range for each county, you will create a legend for this new USA Counties (Generalized) layer.

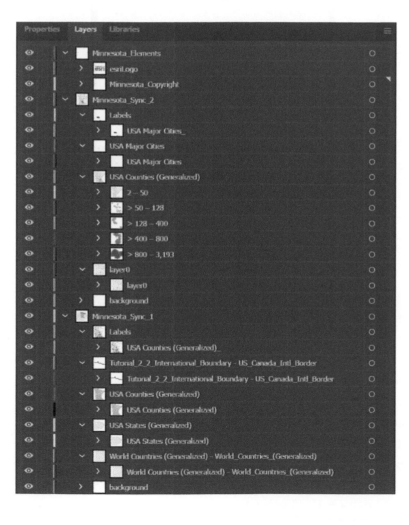

Figure 3.17. The second sync created a new parent layer on the Illustrator file's layer panel, which contains sublayers for the labels and new map layers.

2 In Illustrator, turn off the visibility for all layers except the new USA Counties (Generalized) layer and its sublayer categories (figure 3.18).

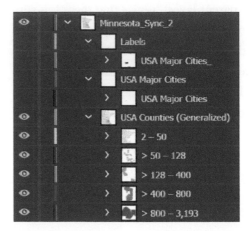

Figure 3.18. Set up your Illustrator layer visibility so that only the USA Counties (Generalized) layer and its sublayers are visible on your map. All visible map feature layers will be included in the legend process.

3 Open the Processes panel by clicking the Processes panel button on the Compilation panel toolbar.

4 Click the Map Legend option from the Processes panel. Then click Create Legend to add a legend to your AI file.

The legend process works by adding a legend item for each visible layer in your AI file. The hierarchy of these items will match that of your layers in Illustrator. When you add a legend with Maps for Adobe Creative Cloud, it will be placed beneath the artboard (figure 3.19), so you may need to zoom out or scroll down to see it. On the Illustrator Layers panel, notice that the new layers containing the legend artwork will be on top of the layer stack.

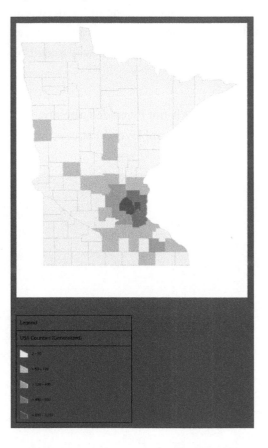

Figure 3.19. The legend added to your map using Maps for Adobe Creative Cloud is placed beneath the artboard.

Layer organization review

Now it is time to inspect your freshly synced Minnesota.ai map and layer structure. Your Illustrator layers for this map should have four main parent layers, from top to bottom: Minnesota_Legend_1, Minnesota_Elements, Minnesota_Sync_2, and Minnesota_Sync_1. Within each of these top-level layers, you will find the following sublayers and artwork:

- **Minnesota_Legend_1**
  - Sublayers for the legend labels, icons, and legend border

- **Minnesota_Elements**
  - An Esri logo, and any data credits available from each sync performed on the Minnesota mapboard

- **Minnesota_Sync_2**
  - Other layer containing a black border fitted to the Illustrator artboard
  - Labels layer with the labels for the major cities
  - Layers for the cities and the second instance of counties added during the second sync, as well as sublayers within the counties for the population density ranges

- **Minnesota_Sync_1**
  - Other layer containing a black border fitted to the artboard
  - Layers with labels for the USA counties

- Sublayers for each of the map layers added in the first sync

Congratulations! You completed both tutorials in this chapter and performed a second sync on the same mapboard. It is now time to complete your map in Illustrator using your own unique style. You can even rearrange layers logically to fit your final desired map. Ultimately, once you have your Illustrator layers set, the design is in your creative hands.

CARTOGRAPHY CORNER

Choropleth maps

A choropleth map visualizes a quantitative attribute associated with geographic areas that are represented by polygons. The hue or color value of the polygon indicates the amount or intensity of the given attribute. Typically, the darker or more saturated a polygon's hue or color value, the greater is the amount of the attribute's value for that area. Conversely, lighter or less saturated colors are typically applied to polygons with a smaller amount or intensity of the attribute's value.

Why normalize choropleth maps?

In most choropleth maps, the geographic areas are not of equal size. In addition to size variation, the mapped attribute may be unevenly distributed or directly associated with an unevenly distributed spatial characteristic. For example, the number of car owners for each US state will be greater in states with a larger population—not because highly populated states attract car owners, but simply because there are more people to own cars. Therefore, a choropleth map showing the raw number of car owners per state would effectively be a population map. Instead, mapping the attribute's rate—or cars per person—is a better method for conveying the spatial story of car ownership. Often, a meaningful rate for choropleth maps can simply be the amount of an attribute per square unit of measurement, which results in a density map, like the population density map of Minnesota that you made in tutorial 3.2 (figure 3.20).

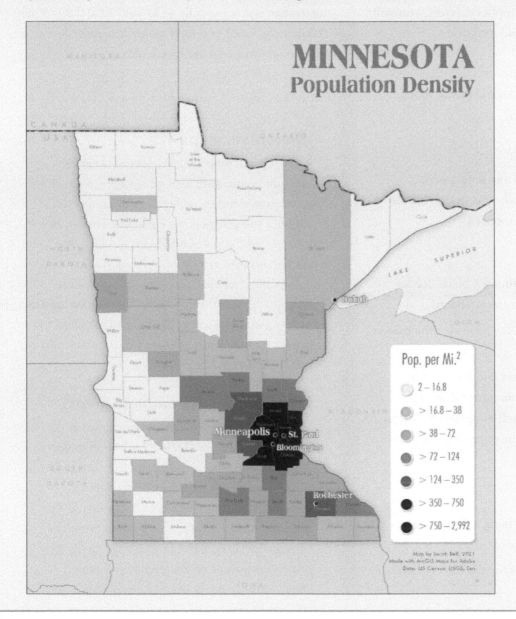

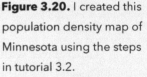

Figure 3.20. I created this population density map of Minnesota using the steps in tutorial 3.2.

CHAPTER 4

MANAGING MAP CONTENT USING MAPS FOR ADOBE CREATIVE CLOUD

IN CHAPTER 2, YOU WERE INTRODUCED TO THE concept of Maps for Adobe Creative Cloud data categories, or libraries. In this chapter, you are going to learn more about these libraries, including how to access your own Maps for Creative Cloud libraries. As you put this new information into practice during this chapter's tutorials, you will also create your own web map using Maps for Adobe Creative Cloud. **Note:** This capability is another premium feature available to ArcGIS Online license users (GIS Professional and Creator only).

Content organization structure and access

Maps for Adobe Creative Cloud organizes ArcGIS Online content into a few different data libraries. In addition to ArcGIS Online content, you can add map layers from your computer. This section describes the various data libraries and how to add local files. Following this discussion on content organization and access, this section concludes with an outline of the steps for organizing your own ArcGIS Online content into Maps for Creative Cloud libraries.

Note that when using Maps for Adobe Creative Cloud with an ArcGIS Enterprise license, the data libraries may differ from what is described in this book, as they will be unique to the Enterprise account used to sign in.

ArcGIS credits

The ArcGIS Online currency is called credits. Because Maps for Adobe Creative Cloud interacts closely with ArcGIS Online, ArcGIS license users can use credits in the extension by syncing premium content, performing certain analytical processes, and adding content to their organization from Maps for Adobe Creative Cloud. For more on ArcGIS credits, go to links.esri.com/ArcGISCredits.

ArcGIS Online data

When creating maps directly from Maps for Adobe Creative Cloud, the most common way to find and add content to a map is by accessing ArcGIS Online data layers from the Compilation panel. If you have been reading along in chronological order, you have already used the Compilation panel to access ArcGIS Online content—or you have at least read about how to do so. Maps for Adobe Creative Cloud catalogs ArcGIS Online data into libraries. All users except for Complimentary accounts have access to the Maps for Creative Cloud, Living Atlas of the World, and ArcGIS Online libraries. These libraries can be found on either the Mapboards panel (Import > From Webmap and Import > From Layer) or the Compilation panel (Add Content > Add Layers and Add Content > Overwrite From Map).

Access to the individual maps and layers in these libraries will vary depending on your user license, as well as the content settings determined by the data publisher (figure 4.1). You can find licensing and required credit information by reading the item description, which can be accessed by clicking an item's thumbnail in the Add Layers window. See "ArcGIS credits" in this chapter for more information.

Data categories for all users

All users have access to the following data libraries, although the content available in each library will vary depending on your license and the item's permission levels:

- **Maps for Creative Cloud:** The Maps for Creative Cloud library contains content specifically curated for Maps for Adobe Creative Cloud use. In this library, you will find ArcGIS Online content that is well suited for building maps in Maps for Adobe Creative Cloud. The Maps for Creative Cloud library catalogs data and layers designed by cartographers specifically for this mapping extension.

- **Living Atlas of the World:** This large collection of map content from ArcGIS Living Atlas of the World has a wide range of thematic options, including the environment, business, health, population, and more.

- **ArcGIS Online:** In this library, you can find publicly available data shared by the Esri Community.

Figure 4.1. The shield icon on the World Traffic Service item, *left*, indicates that it is available to premium subscribers only (Plus, ArcGIS Online, and Enterprise licenses). The target icon with the shield on USA Population Growth, *right*, indicates that this premium content will consume credits.

Data categories for ArcGIS Online and Plus license users

In addition to the three libraries available to all users, ArcGIS Online license users (GIS Professional, Creator, Editor, and Viewer) have access to all the following libraries. Those accessing the extension using a Plus license have access to the My Favorites library.

- **My Organization:** ArcGIS Online users can access their organization's items (image services, feature layers, and more) from this library.

- **My Content:** ArcGIS Online users can access their ArcGIS Online My Content items entries from this library.

- **Groups:** ArcGIS Online users can add content from ArcGIS Groups that they own or are a member of.

- **My Favorites:** Plus and ArcGIS license users have the option to bookmark online content from the Maps for Adobe Creative Cloud Add Layers window. To bookmark an item, click the Favorites button (star icon) on the item's thumbnail image. Bookmarking allows for quick access to your favorite and most used layers.

Local data

All users can add local files to their maps from either the Mapboards panel (Import > From File) or the Compilation panel (Add Content > Add Layer From File). Plus and ArcGIS licenses give access to 4,000 features per file. In tutorial 4.1, you will practice adding local files to a map. See "File types in ArcGIS Pro" in this chapter for information on the different local file types that can be added to a map using the ArcGIS Pro–to–Illustrator workflow.

Sharing maps from Maps for Adobe Creative Cloud to ArcGIS Online

When signed in to Maps for Adobe Creative Cloud using a GIS Professional or Creator ArcGIS Online license, you can share your map compilation from the Compilation panel to your ArcGIS Online account in the form of a web map. These compiled layers can be from local files, ArcGIS Online layers and web maps, or a combination of each. To create a web map from Maps for Adobe Creative Cloud, click the Compilation panel Settings menu button. Then click Share To ArcGIS Online, and provide the required map details on the Share Map dialog box. When sharing from Maps for Adobe Creative Cloud to ArcGIS Online, the map will automatically be added to your My Contents library in the extension. Note that this feature is not part of any tutorials in this book. For more information about creating web maps using Maps for Adobe Creative Cloud, see "Share" in the product documentation at links.esri.com/ShareMaps.

Tutorial 4.1: Making thematic maps

Scenario

In this tutorial, you will be using data from your own computer to make a thematic map in Maps for Adobe Creative Cloud. After you download your synced map as an AI file, you will generate a map legend using the legends process. The first step in this tutorial is to download the assigned data from the web.

Learning objectives

- **Working with local data:** Add data from a local drive to a map from the Compilation panel
- **Working with web maps:** Create a new mapboard from a web map
- **Manually adjusting mapboards:** Freely resize a mapboard extent without constrained proportions
- **Using map projections:** Use the Compilation panel's mapboard setting to update a map's projection
- **Creating graduated point symbols:** Use the Change Style option to create a graduated symbology layer

Working with local data, part 1

Download the CSV file from ArcGIS Online

1. Enter this URL into your browser: **links.esri.com/MappingByDesign.**

2. On the item page, click Download.

3. When the CSV file is finished downloading, save it to your computer in a location that you will remember, such as EsriPress \MappingByDesign.

You will need to access this file during this tutorial.

Working with web maps

Import a web map from the Mapboards panel

1. Open Adobe Illustrator, and sign in to Maps for Adobe Creative Cloud using your ArcGIS Online or Plus account or the ArcGIS trial that comes with this book.

Note that if you are using an ArcGIS Online Viewer or Editor license, some Maps for Adobe Creative Cloud features may not be available to you (see the functionality matrix in table 1.1 in chapter 1 or at links.esri.com /FunctionalityMatrix).

2. On the Mapboards panel, click Import, and then select From WebMap.

3. In the Create From Map window, select the ArcGIS Online library to search from all available ArcGIS Online content.

4. In the Search text field, type **Tutorial_4.1_ WebMap,** and press Enter or Return, and wait for the search to complete.

By default, the results are sorted by highest rating.

5 Click Highest Rated to expose additional options.

6 Select Relevance, and scroll until you see Tutorial_4.1_WebMap.

This item should be at or near the top of the results list (figure 4.2) after sorting by relevance.

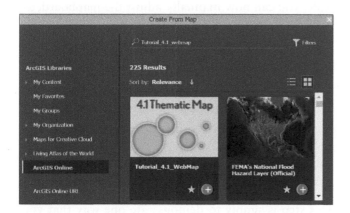

Figure 4.2. The desired web map is shown as the thumbnail on the left.

7 Add Tutorial_4.1_WebMap to your My Favorites library by clicking the Favorites button.

8 Add the web map to the Mapboards panel by clicking the item's Add button, and click Close.

Your Mapboards panel should look like figure 4.3, with a map of the world.

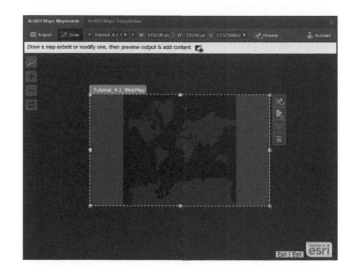

Figure 4.3. Web map added to your mapboard.

Update the mapboard name and extent

When you create a mapboard from a web map, the mapboard is automatically determined by the web map. However, you have the option to adjust the extent and scale of any mapboard before you sync it (see the "What changes can be made in synced mapboards?" sidebar in chapter 3). You will now make some minor adjustments to the mapboard that was automatically created when you added the web map under "Import a web map from the Mapboards panel."

1 Click the Mapboard Options button (second from the top on the vertical Mapboards toolbar) to open the Mapboard Options dialog box.

2 Rename the mapboard **World Cities Population.**

③ Select the button next to Set Level Of Detail so that you can manually update the map scale.

④ Click in the scale text area, and type **100,000,000**.

Your Mapboard Options dialog box should appear as it does in figure 4.4, with the name World Cities Population and a scale of 100,000,000. Your artboard size should be similar to the figure as well, although note that the width and height may differ slightly from the figure, and the units of measurement may be different as well. This is fine.

⑤ Click OK on the Mapboard Options dialog box to apply the updates to the mapboard.

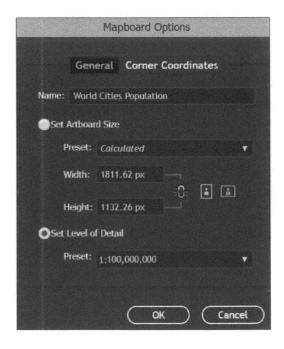

Figure 4.4. Mapboard settings updates to name and extent.

Manually adjusting mapboards

Manually adjust the mapboard extent

Notice in figure 4.3 that the mapboard extent goes beyond the basemap on the left and right sides. Because you selected Set Level Of Detail under "Update the mapboard name and extent" as you updated the map scale, you can now manually adjust the mapboard, as opposed to resizing with constrained proportions.

① Click the mapboard's anchor point on the left side, and drag until that side is snug against the basemap. Repeat for the right side of the mapboard until the mapboard encompasses the entire basemap without too much excess beyond the basemap (figure 4.5).

This step is simply to demonstrate one way that you can manually adjust the mapboard size. The mapboard does not have to be precisely on the edge of the basemap.

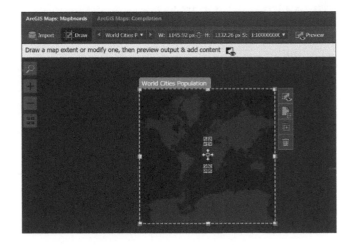

Figure 4.5. The World Cities Population mapboard after manually adjusting to fit the basemap.

Mapping by Design: A Guide to ArcGIS Maps for Adobe Creative Cloud

Using map projections

Reproject the World Cities map

Because Maps for Adobe Creative Cloud allows you to create maps with data hosted on the web, the default map projection that the extension uses is WGS84 Web Mercator (Auxiliary Sphere), commonly known as Web Mercator. Plus and ArcGIS licenses give users access to many other map projections in Maps for Adobe Creative Cloud. In this section, you will be changing the map projection to a more appropriate one for the thematic world map you are creating. To learn more about reprojecting maps using Maps for Adobe Creative Cloud, see "Map projections" in chapter 5.

 Open the Compilation panel, and press Ctrl + 0 (PC) or Cmd + 0 (Mac) to fit the map to the map preview area.

You can also manually resize the panel.

 Delete the basemap.

Click the Current Map Settings button on the Compilation panel toolbar to open the Current Map Settings dialog box.

Click the Projection tab.

In the Search field, type **Winkel**.

This search will return only the map projections with Winkel in their names.

Select World Winkel II from the list of projection results, and click OK.

This projection is primarily used for world maps. When finished, your map should now appear as it does in figure 4.6.

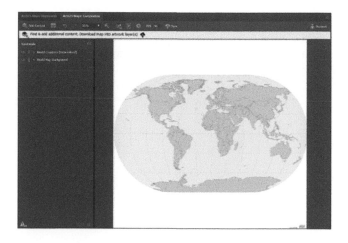

Figure 4.6. The Compilation panel after updating the mapboard projection to World Winkel II.

Working with local data, part 2

Add the downloaded file to the World Cities map

On the Compilation panel, click Add Content, and select Add Layer From File.

Navigate to the location on your computer where you saved World_Cities_Population.csv, and select that file.

Click Open to add the World_Cities_Population.csv file to your mapboard.

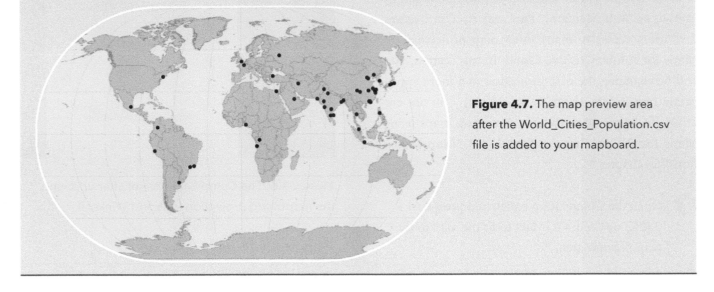
Creating graduated point symbols

Create graduated symbology from the World Cities population attribute

Graduated symbology is a data visualization method used to show the quantity of a feature's attribute. When mapping with this visualization method, a symbol—usually a circular point, although the shape can vary—is placed on the map at the geographic location or geographic center of the object it represents. The size of the symbol is then updated to represent the attribute's quantity for that feature. In this section, you will be using the population attribute of the World_Cities_Population layer to create a graduated symbology map.

1. On the Contents panel, click the World_Cities_Population Options button. Then select Change Style.

2. On the Style panel, click the drop-down arrow under Choose An Attribute To Show, and select Population_2020 from the list.

This numerical selection will automatically convert the symbols in this layer to graduated symbology, and natural breaks will be the default classification.

Customize the default graduated symbols

1 Click Options from Select A Drawing Style in the Counts And Amounts (Size) section (figure 4.8).

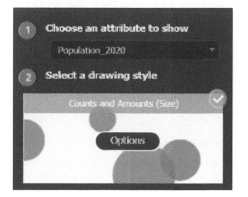

Figure 4.8. Clicking the Options button will expose more features for symbolizing this layer.

2 On the Style panel, update the class breaks to **6** so that there are six classification groups for this layer.

In the histogram, keep the uppermost and lowermost values as is for now (37,393,129 and 7,220,104, respectively).

3 Update the five remaining middle values by clicking the current values and updating them top to bottom as follows:

- **30,000,000**
- **20,000,000**
- **15,000,000**
- **10,000,000**
- **8,000,000**

Next, you will round the largest and smallest values of the histogram by manually updating the values at the top and bottom axes.

4 Click the top value (37,393,129), and type **37,500,000** to round this value. Do the same for the bottom value, this time rounding down by typing **7,200,000**.

5 Change the maximum size for your graduated symbols by clicking the maximum size text field's value, and type **60**, making the largest symbols on your map 60 pixels in diameter. Then change the minimum symbol size to **5** pixels.

6 Click OK to save the new style.

The style (figure 4.9) will only be applied by clicking OK.

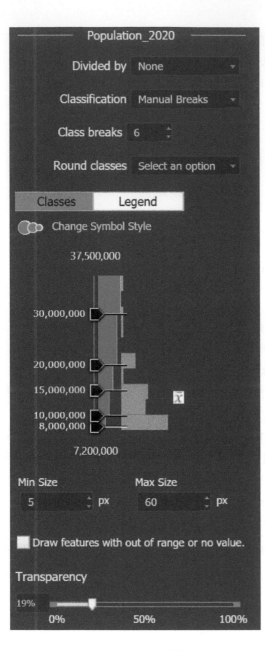

Figure 4.9. The Style panel filled in for six manual class breaks.

Add city labels

① Show the World_Cities_Population layer options, and select Manage Labels from the list of options.

This setting will automatically add labels to the cities layer using the City_Name attribute. Although these labels appear on the map, they are not yet added to your map until this section is completed.

② Change the text size to **8,** and then change the font face from bold to regular by clicking the Bold button in the label options.

③ In the Alignment picker, select the center option so that the cities' labels are directly centered on their symbol.

④ Check the Show Overlapping Labels box (figure 4.10) so that all labels are visible on the map.

⑤ Click OK to add the labels to the map.

Figure 4.10. The World_Cities_Population layer's label settings.

Checkpoint

At this point in the tutorial, the Compilation panel (figure 4.11) should have three layers:

- World_Cities_Population
- World Countries (Generalized)
- World Map Background (oceans)

The city points should be symbolized as graduated points, and they should all be labeled by city name.

Figure 4.11. The completed Compilation panel with city points graduated and labeled.

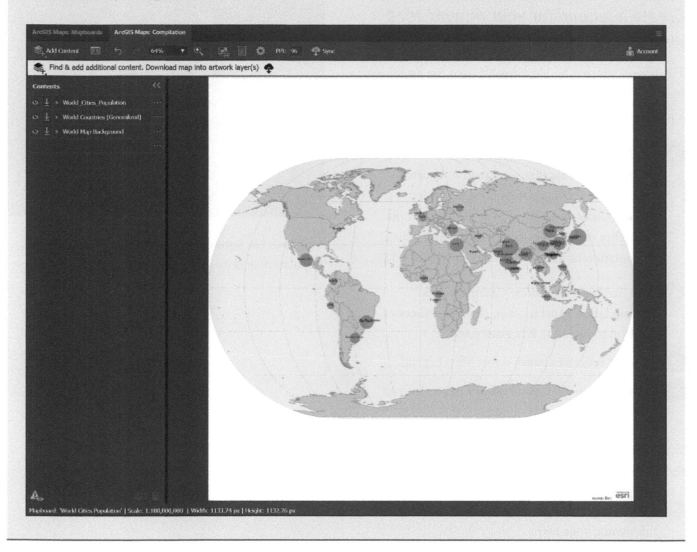

Sync the World Cities Population map

Your World Cities Population thematic map is now ready to be downloaded.

1 Click Sync on the Compilation panel toolbar, and wait for your map to be built in Illustrator.

Create a legend for your downloaded World Cities Population map

The legend process builds a legend item for each visible layer in an AI file. In this step, you will be creating a legend for the cities' graduated symbols layer.

1 In Illustrator, investigate the World Cities Population.ai file that you created. Locate the World_Cities_Population layer under World Cities Population_Sync_1. Turn off the visibility of all layers except the World_Cities_Population layer and its sublayers (figure 4.12).

2 On the Maps for Adobe Creative Cloud Compilation panel toolbar, click the Processes button to open the Processes panel.

3 On the Processes panel, select Map Legend.

4 Click the Create Legend button to create your legend.

When the legend is generated, it will be located below the artboard in Illustrator.

5 Turn layers back on as needed to finish designing your map.

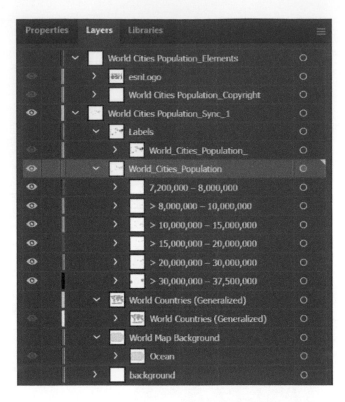

Figure 4.12. Setting visibility for only World_Cities_ Population and its sublayers.

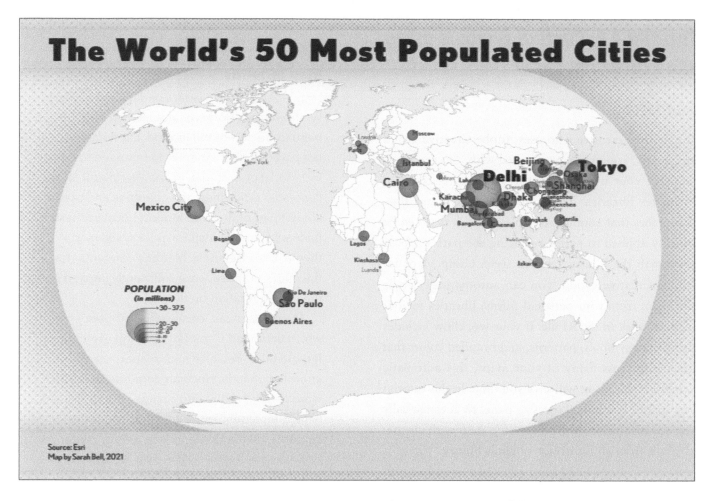

Figure 4.13. I created this graduated symbology map following the steps in tutorial 4.1, and then styled it in Illustrator.

Tutorial 4.2: Custom symbols using the Processes panel

Scenario

If you are a creative who uses Adobe Creative Cloud frequently, you may have worked with, or even made some of your own, Illustrator Symbols, Brushes, or Swatches libraries. These Adobe libraries are Illustrator files that contain reusable artwork that can be quickly applied to paths, replacing the path's appearance with the library artwork style. Using Maps for Adobe Creative Cloud, you can automatically apply the styles from your personal Adobe libraries to your map artwork in the AI file. If your workflow includes using map symbols, patterns, and branded colors that are used across many of your maps, this automatic symbol replacement will save you time. In this tutorial, you will be using the Processes panel to automatically replace default point symbols on a map with custom symbols from an Illustrator symbols library.

Learning objectives

- **Working with web maps on the Compilation panel:** Populate an existing mapboard with a web map on the Compilation panel

- **Using Processes panel symbolization:** Use automatic symbolization with custom Illustrator symbols

Map legends

Map legends help readers understand a map's symbolization. However, not all map symbols need to be part of the legend. Elements that are not part of a map's theme do not need to be included. For example, in *The World's 50 Most Populated Cities* map, the countries and ocean are not part of the map legend because these features are not a major aspect of the map's theme. Likewise, these features' referents—for the map's purposes—are sufficiently implied by their symbolization on the map.

Mapmakers can increase the efficiency by which their map is read by creating useful legend titles. Because map readers usually know that map legends are legends, using the word *Legend* as the title isn't necessary. Instead, aiming for a descriptive title that succinctly captures the map's theme can reduce clutter and increase legibility. The map legend title in contest winner Andrew Bernish's *Public Kayak Launches of Eastern Talbot County, Maryland* (figure 4.14) is the actual title of the map as well. How's that for efficiency? The title tells you that this is a map of kayak launches, so Bernish does not even need to put the kayak launch symbol in the legend. There is one single symbol: the shape of Eastern Talbot County, which is symbolized by the same bright-yellow color used for the symbol referent.

Figure 4.14. *Public Kayak Launches of Eastern Talbot County, Maryland* by Andrew Bernish.

Download the data

This tutorial requires an Illustrator symbols library called Oslo_Symbols.ai. First, you will download this file and save it to your computer.

 Enter this URL into your browser, **links.esri.com/MappingByDesign**.

2 From the account's item page, click Oslo_Symbols.

3 On the item page, click Download.

4 Once the file is finished downloading, save it to your computer in a location that you will remember, such as EsriPress\MappingByDesign.

You will be accessing this symbols library for this tutorial.

Working with web maps on the Compilation panel

Draw a large-scale mapboard

In this section, you will draw a large-scale map over the Sentrum neighborhood in Oslo, Norway.

1 Open Illustrator, and sign in to Maps for Adobe Creative Cloud.

2 On the Mapboards panel, click the Search button on the basemap.

3 In the Search text area, type **Sentrum, Norway**, and click the first option in the results list to center the map on the Sentrum region of Oslo, Norway.

4 Use the basemap selector to select a basemap of your choice.

This tutorial uses the Streets basemap.

5 Click Draw, and create an extent like that in figure 4.15.

You may need to zoom in to see the same amount of detail as shown in the figure.

Figure 4.15. The approximate mapboard extent ideal for tutorial 4.2.

Update your mapboard settings

To complete your large-scale mapboard for Sentrum, you will be updating the settings in the Mapboard Options window.

1 On the Mapboard Options dialog box, name your mapboard **Sentrum,** and click Set Level Of Detail to set the map scale to **1:4,500** (figure 4.16).

2 Click OK.

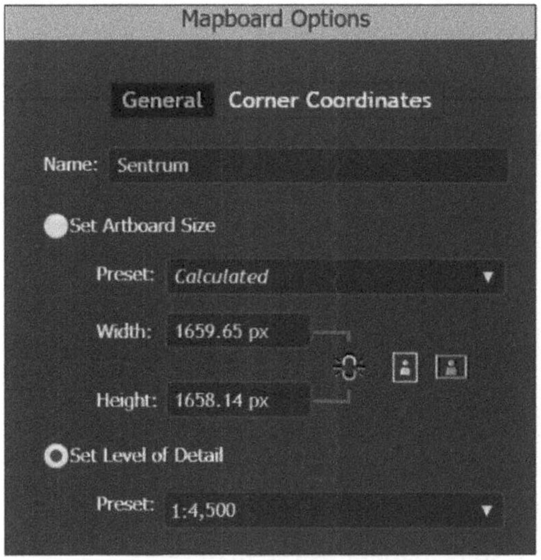

Figure 4.16. The Sentrum mapboard settings. The unit of measurement may be different for you.

Add a web map from the Compilation panel

In tutorial 4.1, you generated a mapboard automatically by adding a web map on the Mapboards panel. In tutorial 4.2, you will be adding a web map to your manually created Sentrum mapboard, this time on the Compilation panel.

1 Open the Compilation panel, and press Ctrl + 0 (PC) or Cmd + 0 (Mac) to fit the map to the map preview area.

2 Click Add Content, and then select Overwrite From Map.

3 Select the ArcGIS Online library, and search for **tutorial_4.2_sentrum.**

4 Click the Add button to add the web map that appears (figure 4.17).

An alert will appear, letting you know that by adding a web map versus a feature layer, any data that may have already been added will be removed.

5 Click OK to accept, and close the Overwrite From Map panel.

> **Reminder**
>
> You can add non-web map layers after adding a web map.

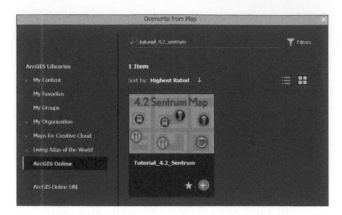

Figure 4.17. The web map is added by clicking the Add button for the 4.2 Sentrum Map option after searching for Tutorial_4.2_Sentrum.

Note: If you are signed in with a Plus account or if you have not upgraded to Maps for Adobe Creative Cloud 3.0, the vector tile basemap in the Tutorial_4.2_Sentrum web map will not be available to you. If this is the case, select the VectorStreetMap basemap from the basemap selector on the Contents panel.

Style points by a categorical attribute

You will soon be automatically replacing the points of interest (POI) layer's default symbology with custom symbols from the Oslo_symbols.ai library. For the symbols from the Oslo_symbols.ai library to match up correctly to your map's POI layer, you will first need to symbolize these points by category.

1 From the osm_points – point Layer Options menu, click Change Style.

2 On the Style panel, from the Choose An Attribute To Show drop-down list, select the Amenity attribute.

3 Click OK.

Selecting the Amenity attribute applies a unique color to each type of amenity in this layer.

Add labels to the POI layer

1 Reveal the layer options for the osm_points – point layer again, and this time select Manage Labels.

2 On the Label panel, check the box next to Show Overlapping Labels so that all labels from the layer will appear, and then click OK to add the labels to your map (figure 4.18).

Sync to download the Oslo map

1 Click Sync to download the map to Illustrator.

The map you have just compiled contains a lot of detailed data, so be patient as the map syncs.

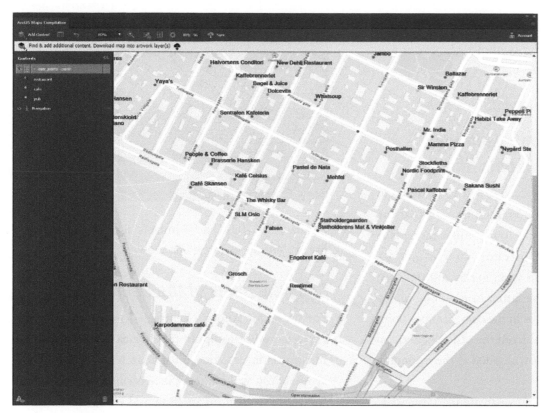
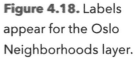

Figure 4.18. Labels appear for the Oslo Neighborhoods layer.

Checkpoint

You now have a new Sentrum.ai file that contains the following layers:

- **Mapboard_Elements:** Parent layer containing a vector Esri logo and any credit information from the map layers

- **Mapboard_Sync_1:** Parent layer containing all layers added during the first sync

- **Labels:** Layer containing sublayers for all labels. Because you added labels to only one layer, you will have just one sublayer named osm_points - point.

- **osm_points - point:** Layer containing sublayers for each Amenity category.

  - restaurant

  - café

  - pub

- **World Navigation Map:** The vector tile basemap parent layer, which contains sublayers for all vector data and labels from this basemap. If you selected a different basemap, this layer will reflect that selection.

- **background:** Layer containing a white rectangle

Next, you will move onto part 2 of this tutorial, where you will be using the symbol replacement process on an already synced AI file.

Using Processes panel symbolization

Identify the four symbols in the Oslo_Symbols.ai symbols library

In this section of tutorial 4.2, you will be working in the Sentrum.ai file that you created in part 1. Using Maps for Adobe Creative Cloud, you can replace default artwork with an Illustrator library's artwork either during or after a map sync; in this tutorial's second part, you are doing the latter. You will be replacing the default point symbols in the osm_points – point layer with a custom symbol from the Oslo_Symbols.ai symbols library that you downloaded at the beginning of this tutorial.

Before you begin part 2, investigate the layer structure of your Sentrum.ai file. You will find the osm_ points – point layer and its categories just beneath the Labels parent layer's categories.

1 In Illustrator, navigate to the location where you saved the Oslo_Symbols.ai file, and open it.

2 Open the Illustrator Symbols panel to view this library's symbols (Window > Symbols).

3 Using the Symbols panel menu (upper-right button consisting of three horizontal lines), select Large List View.

You should see the following four symbols: BUS, CAFE, BEER, and RESTAURANT (figure 4.19).

Figure 4.19. The Oslo_Symbols.ai symbols library list of symbols in Large List View.

Copy the name of the BEER symbol

For Maps for Adobe Creative Cloud to identify the correct symbol to apply to your map's artwork, the Illustrator layer name must match the name of the symbol. You will rename the Sentrum.ai layers so that they are identical, in spelling and case, to the names of the symbols in the Oslo_Symbols library. You can copy the name directly from the Symbols panel, or you can just remember that the name of the pub symbol is BEER (case sensitive).

1 With the symbol name BEER either copied or remembered, open or return to the Sentrum.ai file that you created in part 1.

2 Click the text of the osm_points – point layer's pub sublayer in Illustrator to make the layer name editable, and change it to **BEER** by either pasting it from the clipboard or typing it.

3　Referring to the Oslo_Symbols.ai library, rename the remaining osm_points – point sublayers in your Sentrum.ai file to match their corresponding symbol names:

- restaurant to **RESTAURANT**
- cafe to **CAFE**

Select the symbols library from the Processes panel

1　Keeping your Sentrum.ai file as the active Illustrator file, open the Maps for Adobe Processes panel.

You can do this by clicking the Processes button on the Compilation panel toolbar.

2　On the Processes panel, click Symbols to activate that process.

3　Click Set Path, and then navigate to the Oslo_Symbols.ai file on your machine and click Open.

4　Make sure that Run On Sync is not checked, since you are applying this process to the AI file that you already created.

5　Click Replace Symbols to apply the process.

This will take a few moments.

Once the Custom Symbols process is completed, your map should appear similar to figure 4.20. There may be some variation in the background data layers.

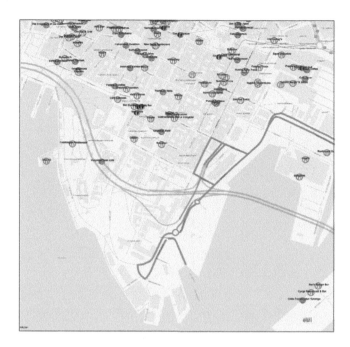

Figure 4.20. The map of Sentrum points of interest after performing all the steps in tutorial 4.2.

6　Style your map in Illustrator as you wish.

Now that you have completed all the steps in tutorial 4.2, you are free to apply your own unique design using Illustrator (figure 4.21).

Figure 4.21. I created this map of Sentrum POI in Oslo by following the steps in tutorial 4.2, and then further designed it in Illustrator.

Mapping by Design: A Guide to ArcGIS Maps for Adobe Creative Cloud

Local data

Maps for Adobe Creative Cloud connects Illustrator to the power of ArcGIS, which means that users have access to many ArcGIS Online mapping features. One of the many features available in ArcGIS Online is the ArcGIS World Geocoding Service. This geocoding service calculates the appropriate spatial coordinates for features and correctly positions them on maps. For example, in tutorial 4.1 you added the World_Cities_Population.csv file from your local files to the mapboard. The correct placement of these comma-separated value (CSV) data was possible because the ArcGIS World Geocoding Service read the City_Name and Country fields, and then created a geographic point on the map for each city's record. Because the geocoding service is automatic, the city points were instantly placed on their respective city locations on the mapboard. This section outlines the local file types that can be added to a Maps for Adobe Creative Cloud mapboard.

Shapefiles

A shapefile is a vector data storage format developed by Esri. A single shapefile comprises a set of related files. To add a shapefile to a mapboard, the related files must all be placed in a single ZIP (.zip) folder. Then select the ZIP folder from either the Mapboards panel (Import > From File) or from the Compilation panel (Add Content > Add Layer From File).

Note about shapefiles

Among the several different related file types that a shapefile can consist of, the following four file types must be present in the ZIP folder when adding a shapefile to a mapboard: SHP, SHX, DBF, and PRJ.

CSV and TXT files

Text files (TXT), as their name implies, are electronic text. A text file that uses a specific delimiter to separate its text as values can be read as a tabular dataset. Much like TXT files, CSV files use commas to delimit values. Programs that parse TXT and CSV files into tabular format will organize the table's first row as the fields (attributes) and the remaining rows as the records (individual instances or features). In the World_Cities_Population.csv file, the fields are CityRank, City_Name, Country, and Population. The records are the rows of values for these fields. Take the most populated city in this CSV file, which is Tokyo, as an example. The record representing Tokyo is its entire row (figure 4.22), in which the fields' values are 1, Tokyo, Japan, and 37393129 (the city's population for 2020).

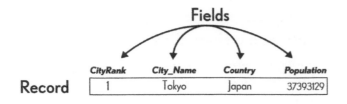

Fields

CityRank	City_Name	Country	Population
1	Tokyo	Japan	37393129

Record

Figure 4.22. The fields in this image are CityRank, City_Name, Country, and Population, and the record is the entire row's values for those fields—1, Tokyo, Japan, 37393129.

MAPMAKER TIP

Making TXT files for use in Maps for Adobe Creative Cloud

ArcGIS can parse TXT files that use a semicolon, comma, or tab to separate fields. TXT files with other separators cannot be used in Maps for Adobe Creative Cloud.

To add a CSV or TXT file from your computer, the file must include an appropriate location field, and the first row in the file must contain the location field names. The location field can be as general as a city name or as precise as latitude and longitude coordinates. Coordinates must be separated into two fields: one for the x-coordinate and another for the y-coordinate. For a complete list of what location fields are supported by ArcGIS, go to links.esri.com/Spreadsheet.

KML and KMZ files

Keyhole Markup Language (KML) files are tag-formatted files that follow the Extensible Markup Language (XML) formatting standard. A compressed KML file is called a KMZ file. Location fields must be stored as tags to be added to a map. For more information on using KML and KMZ files in ArcGIS, go to links.esri.com/KML.

GPX files

GPX files are Global Positioning System (GPS) exchange files that use an XML format. Because GPX files are created from GPS-recorded waypoints, the map features are points or lines. See "CVX, TXT, and GSP Files" in the ArcGIS Online help for more details on using GPX files in the ArcGIS Online environment.

File types in ArcGIS Pro

ArcGIS Pro supports a wide variety of data formats that go beyond the file types listed in this section. When these data are added to an ArcGIS Pro map and then shared as an AIX file, you can use Maps for Adobe Creative Cloud to open the AIX file in Illustrator. The respective map features will appear as they did in ArcGIS Pro. To learn about the file types that are supported by ArcGIS Pro, see "Supported data types and items" at links.esri.com/DataTypes.

CHAPTER 5

ADVANCED MAPMAKING USING MAPS FOR ADOBE CREATIVE CLOUD

Maps for Adobe Creative Cloud provides many cartographic and GIS analysis—or geo-analysis—tools that can help mapmakers uncover spatial relationships that may exist within data. This chapter is dedicated to working with the advanced Maps for Adobe Creative Cloud geo-analysis tools that can be accessed on the Compilation panel.

Using analytical tools for better map stories

Although this book does not contain a deep dive into the science and math of GIS itself, it is a book for every mapmaker who wants to use Maps for Adobe Creative Cloud. To speak to this broad audience of mapmakers, this chapter touches on some of the concepts behind these geo-analysis tools. **Note:** For full access to these geo-analysis tools, a Plus or Arc-GIS Online (GIS Professional or Creator) license is required. ArcGIS Enterprise users with the appropriate portal settings can also use these tools. For more information, see "Configure general settings," at links.esri.com/PortalSettings. For the tutorials in this chapter, it is recommended that you use a license type that grants full access to the geo-analytical tools or use the free trial provided with this book.

Map projections

Earth is an ellipsoid, in which the diameter at the equator is slightly larger than the diameter between the poles. Map projections, put simply, are the geometrical transformations that allow us to portray our ellipsoid-shaped Earth's surface on a flat map. Each map projection is designed to preserve a specific spatial property while unavoidably distorting others. Do not let this distortion discourage you. From the time humans first began making maps, distortion has been necessarily woven into the fabric of cartography. Transforming a curved surface of an ellipsoid into flat planar representation requires distortion. So, you need to know which spatial property you want to preserve when choosing a map projection. By knowing the properties that become distorted with a projection, you can also understand which properties are preserved. In general, map projections can preserve local angular relationships, distances, areas, and directions.

Many projections focus on preserving one of these characteristics, resulting in—sometimes drastic—distortion among the remaining characteristics. However, some projections are developed for the purpose of striking a compromise of distortion across more than one characteristic, thereby not fully preserving any single characteristic but rather balancing and minimizing the distortion among each of the characteristics.

As stated in this chapter's introduction, diving into the science behind these projections would detract from this book's purpose: serving as a user guide for Maps for Adobe Creative Cloud. Still, because Maps for Adobe Creative Cloud users can choose from so many map projections, it is good to equip you with a basic knowledge of how projections work. No matter which projection you choose, when changing a map's projection—a process known as reprojection—there are certain steps to follow in Maps for Adobe Creative Cloud. If you have been reading this book in linear fashion, you have already performed steps to reproject a map in tutorial 4.1, but you have not yet been presented with the reasons behind those steps. The following section elucidates the Maps for Adobe Creative Cloud reprojection workflow.

Web Mercator

A huge advantage of Maps for Adobe Creative Cloud is its connection to ArcGIS Online, Esri's web mapping software as a service. As discussed in chapter 4, because of this direct access to the web, Maps for Adobe Creative Cloud uses WGS84 Web Mercator (Web Mercator) as the default projected coordinate system. Web Mercator is the most widely used default projected coordinate system in web-based mapping applications. Because of its ubiquitous presence, most people who use interactive and navigation maps on the web are unwittingly accustomed to reading maps in Web Mercator.

Web Mercator is based on the Mercator projection, which dates to at least 1569 when cartographer Gerardus Mercator created maps for sea navigation. It makes sense then that the Mercator projection preserves angles and compass bearings since that is what is needed for navigation. Thus, area and distance are necessarily distorted to compensate for preserving angles. Today's modified web-based version of Mercator, Web Mercator, departs slightly from its predecessor by also distorting angles and compass bearings together with its more drastic distortion of area and distance.

Although Web Mercator is widely used, its distortions are not broadly known across its user base. In figure 5.1, all the yellow ellipses on both maps cover the same amount of Earth. These ellipses are called Tissot's indicatrix after French mathematician Nicolas Auguste Tissot, who introduced these ellipses to represent the spatial pattern of map projection distortion. In figure 5.1, all ellipses on both maps are placed in identical locations—at 30-degree parallels and meridians. The map on the left represents Web Mercator, and on the right is a projection known as the World from Space, intended to show Earth as if it were viewed from space. Thus, not all of the planet is intended to be visible in this projection.

Each ellipse covers an identically sized surface area. Knowing this, you can see how the Web Mercator map, on the left, distorts area.

Figure 5.1. A comparison of Tissot's indicatrix over two map projections. Each ellipse in both maps covers the same amount of surface area. The Web Mercator map on the left shows how this projection distorts objects as they move farther from the equator, making objects appear larger than they really are. The map on the right uses the World from Space projection.

Important steps for reprojecting maps in Maps for Adobe Creative Cloud

Maps for Adobe Creative Cloud requires a specific order of operations for map reprojection. You may recall from tutorial 4.1 that a map is reprojected from the Compilation panel. This section includes additional important information regarding map projection.

Add raster data after reprojecting a map

If you are using Maps for Adobe Creative Cloud to change your map projection, it is advised to add raster data after your map is reprojected. You can add from among many different raster layers after the map is reprojected, but during reprojection certain raster layer types will be automatically removed from the Compilation panel. If this occurs, you will receive the following message:

ERROR: The layer [layer name] cannot be added to the map. Layer's spatial reference is different from the map.

So again, if you want to add raster data to a map that you are planning to reproject, first reproject the map, and then add the raster data.

Change PPI settings before adding raster data

The default raster resolution in Maps for Adobe Creative Cloud is 96 pixels per inch (PPI) and can be manually changed up to a PPI of 300. If you are choosing a map projection other than Web Mercator, it is recommended that you change the PPI settings before adding raster layers. If you update the PPI after adding raster layers to a mapboard that has already been reprojected, it will give you the same error described in the previous paragraph.

Geo-analysis tools

Until now, you have been using the Contents panel Layer Options menu to change a layer's style, add labels and data filters, and bookmark layers as favorites. While you were accessing the menu, you probably noticed the advanced geo-analyses options available there: Visualize Route, Create Buffers, Visualize Travel Times, and Add Demographic Data (figure 5.2). You are about to explore these data-driven geo-analyses in this chapter's tutorials.

Figure 5.2. The Layer Options menu includes filters, labels, and geo-visualization. Not all options are available for each data type.

Among the geo-analyses options that you will learn in this chapter, the operations may function differently based on their type. In fact, some geo-analyses will work only on specific data types. The following section will help clarify how and when to use geo-analyses functions available on the Compilation panel's Contents panel.

Note: Full access to these features requires a Plus or ArcGIS Online (GIS Professional or Creator) license or the free trial provided with this book. Enterprise users with the appropriate portal settings can also access these tools.

Visualize Route

The Visualize Route feature allows you to connect pairs of points and measure the distance or time between the pairs. This operation results in a new layer of route lines that connect origin and destination points. You have the option to create straight lines or create lines based on mode of travel, such as the walking path or driving route between the start and end points.

The Visualize Route geo-analysis function requires two input layers: one origin point layer and one destination point layer. Both layers must have an identical number of records, or at least one of the two layers must have one record only. Neither input layer can have more than 5,000 points. The more points in a layer, the longer is the processing time. The Visualize Route geo-analysis will produce a new layer on the Contents panel containing the route, which will be lines. These lines will have attribute information such as distance and driving time between the connected points; these attributes can be used for labeling.

Create Buffers

With this tool, you can create geographic buffers of specified distances around point, line, and polygon features. A new layer containing buffers as polygons is created around the selected input layer. Buffers will be created only around input features that are within the mapboard extent.

Visualize Travel Times

With this operation, a new polygon layer is created showing the areas that can be reached within specified travel times or travel distances. The input layer for the Visual Travel Times geo-analysis tool should be a single layer of points with no more than 1,000 features.

Add Demographic Data

The Add Demographic Data operation provides demographic and landscape information related to people, places, and boundaries that are associated with the input point, line, or polygon layer. The operation uses the ArcGIS® GeoEnrichment℠ service. To learn more about this service, see "Enrich Layer" in the ArcGIS Online help.

Tutorial 5.1: Winnipeg airport map

Scenario

The purpose of this tutorial is to familiarize you with the Create Buffers and Visualize Travel Times geo-analysis operations, as well as to demonstrate the resulting Illustrator layer structure for each. The tutorial is divided into two parts: one for each geo-analysis operation. Because this is a learning exercise for understanding the function of these geo-analyses, you will not be presented with all the steps to construct a complete map. You have learned about adding contextual layers, labeling, and syncing in previous chapters. After completing this tutorial, you are encouraged to explore adding new layers, labels, and other elements that you may want to use to complete this map.

Learning objectives

Geo-analysis, part 1

- **Creating buffers:** Add geographic buffers to a map using the Create Buffers geo-analysis tool
- **Mapboard preset sizes:** Define a mapboard's output size using the preset options

Geo-analysis, part 2

- **Visualizing travel times and distances:** Visualize the travel distance area that can be reached within specific time frames
- **Layer organization:** Inspect the resulting Illustrator layer structure from the geographic buffers and Visualize Travel Times geo-analyses

Creating buffers

Draw a mapboard around Winnipeg, Manitoba, Canada

For this tutorial, you must use a Plus or ArcGIS Online (GIS Professional or Creator) license. You may also use the ArcGIS Online trial license offered with this book.

Note: The layer structure for Plus users may differ slightly from the figures in this tutorial.

1. Sign in to Maps for Adobe Creative Cloud.

2. On the Mapboards panel, click the magnifying glass to open the search, and type **Winnipeg** in the search area. Select the first option listed in the search results, Winnipeg, Manitoba, CAN.

③ Use the basemap selector to switch the basemap to a basemap of your choice.

④ Zoom in or out so that your map area looks like figure 5.3, centered on Winnipeg so that the east and west edges of the mapboard are about 40–60 miles from the city, and the north and south edges are roughly 30–50 miles from the city.

To easily fit the map, you can adjust the size of the Mapboards panel before drawing the mapboard.

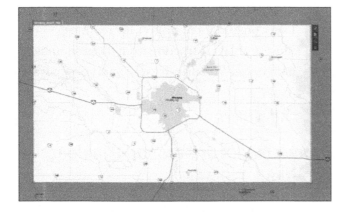

Figure 5.3. Winnipeg mapboard extent as viewed on the Mapboards panel.

⑤ Once your map preview area is similar to that shown in figure 5.3, click the Draw mapboard tool, and then draw a mapboard as similar to the mapboard in the figure as possible.

Mapboard preset sizes

Update the mapboard properties

Until now, your mapboard extents have been defined either using the Draw tool or by adding a web map. After adding a mapboard, you can also adjust it by choosing from common preset sizes. In this section, you will update the mapboard size using one of these presets.

① Click the Mapboard Properties option button on the mapboard toolbar, second button from the top on the vertical toolbar.

② In the Name text field, type **Winnipeg_Airport_Map**.

③ Under Set Artboard Size, from the preset drop-down list under Web, select Web-Large (1920 × 1080 px).

The units of measurement (pixels, points, inches, centimeters, and so on) are determined by your settings in Illustrator; therefore, the units may differ from the figure. By choosing Web-Large, the size will be 1920 × 1080 pixels, even if the units are expressed differently.

After Web-Large (1920 × 1080 px) is selected, the map scale should be approximately within the range of 1:275,000 to 1:295,000 (figure 5.4). Your map scale may differ slightly, which is fine. If it is drastically different—if the denominator is smaller than ~200,000 or larger than ~300,000—you may want to start again at section 1, "Draw a mapboard around Winnipeg, Manitoba, Canada," as it could impact how the geo-analyses results appear in your map.

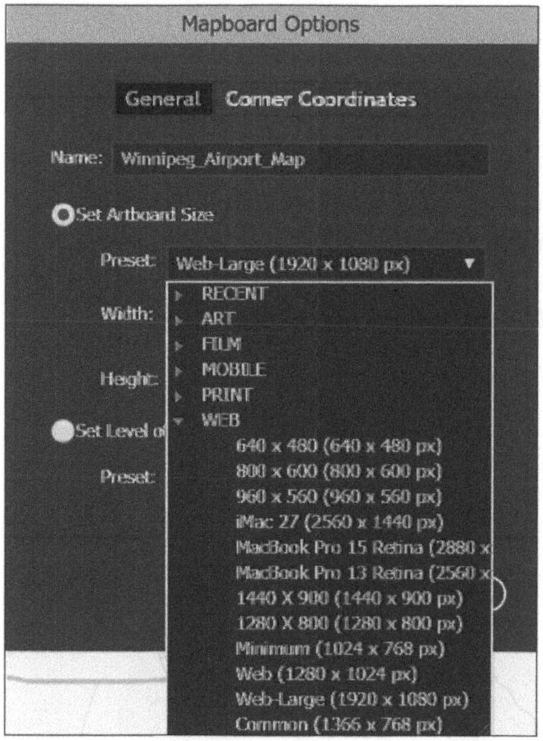

Figure 5.4. The mapboard output size is Web-Large, and it is named Winnipeg_Airport_Map.

4 Click the Draw tool button to disable it.

Add content to the Winnipeg_Airport_Map mapboard

1 Click the Preview button to open the Compilation panel.

2 Click Add Content, and select Add Places from the menu options. In the Search text field, type **YWG Airport,** and select YWG, 2000 Wellington Ave, Winnipeg, Manitoba, R3H, CAN.

After selecting this option, you will get a small list of results.

3 From these results, click the Add button (+) for YWG – 2000 Wellington Ave, Winnipeg to add the Winnipeg airport point to your map.

4 Close the Add Places window.

Reminder

You can see a layer's options by hovering the pointer over a layer's Options button.

Add geographic buffers

1 On the Contents panel, expose the layer options for the YWG – 2000 Wellington Ave, Winnipeg layer that you just added. Then select Create Buffers from the options.

The Create Buffers dialog box opens.

You will update the settings on the Create Buffers dialog box so that the operation will generate three buffers that extend 5, 10, and 20 kilometers from the YWG Winnipeg airport.

2 In the buffer size text area, type **5 10 20** with a single space between each set of numbers, and then select kilometers as the units (figure 5.5).

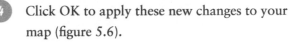

Create buffers from **YWG 2000 Wellington Ave**

1 Enter buffer size

5 10 20   Kilometers

To create multiple buffers, enter distances separated by spaces (2 3 5).

2 Result layer name

Buffer of YWG 2000 Wellington Ave (5 10 20 Kilometers)

Apply

Figure 5.5. Create Buffers dialog box is set to create 5-, 10-, and 20-kilometer buffer rings around the YWG Airport in Winnipeg, Manitoba.

3 Click Apply, and wait for the geo-analysis to complete.

The geo-analysis may take a few moments.

Once the geo-analysis is complete, you will see the drive time polygons in the map preview area. Although you can preview these polygons, they will not be saved to your mapboard until you click OK on the Create Buffers dialog box.

4 Click OK to apply these new changes to your map (figure 5.6).

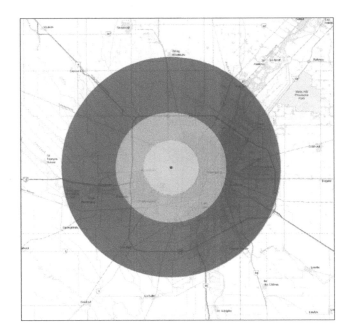

Figure 5.6. The map preview area after creating multiple geographic buffers around the airport.

Visualizing travel times and distances

Visualize driving time from the airport

In Maps for Adobe Creative Cloud, the Visualize Travel Times and Visualize Travel Distances operations work similarly. Both operations are accessed from the Visualize Travel Times option in a point layer's Layer Options menu. When choosing this option, you can select from travel times or distances from the measure drop-down list (see figure 5.7 where Driving Time is selected).

In this part of tutorial 5.1, you will be adding polygons to visualize areas that can be reached by driving within 10, 15, and 20 minutes from the YWG Airport in Winnipeg, Manitoba.

1. Expose the YWG 2000 Wellington Ave layer options, and select Visualize Travel Times.

2. Keep the measure as Driving Time, and enter **10 15 20** in the text area next to minutes. Keep the unit of time in minutes.

3. Click Apply to begin the geo-analysis operation.

The geoprocessing will take a few moments.

4. Once the operation is complete, click OK to add these drive time polygons to your map.

The result will look like figure 5.8.

Figure 5.7. The Visualize Travel Times settings will produce 10-, 15-, and 20-minute drive time polygons from the YWG airport in Winnipeg, Manitoba.

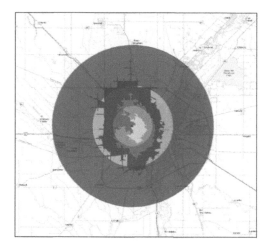

Figure 5.8. The map preview area after adding drive time polygons to the Winnipeg_Airport_Map.

Layer organization

Sync your Winnipeg_Airport_Map.ai layers

 Inspect the Winnipeg_Airport_Map.ai layers.

Examine the Illustrator layers in your newly created AI file to learn about the resulting layer structure for the Create Buffers and Visualize Travel Times geo-analysis operations. In Illustrator, the layers in your Winnipeg_Airport_Map.ai file should contain the layers that you see in figure 5.9.

Geographic buffers layer

In the AI file, you will find a parent layer for the geographic buffers that you created in part 1 and sublayers within this parent layer that contain polygons for the individual buffer rings. Because you performed the Create Buffers operation on a point layer that had only a single point, you will have only one buffer ring per sublayer category. When the Create Buffers operation is performed on a layer that contains multiple points, there will be as many rings in these sublayers as there are points.

Travel time layers

You will notice a parent for the travel time polygons. Notice that within this parent layer, there are three sublayers containing the travel time areas that you generated in part 2 of this tutorial. There is a polygon in each of these sublayers for every point on which the travel time operation is performed. Because you used only a single-point layer that contained the YWG Airport in Winnipeg, each of these sublayers will contain only one polygon.

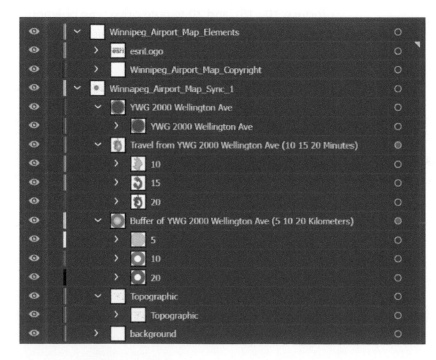

Figure 5.9. After the Winnipeg_Airport_Map is synced, these layers will be present in your newly created AI file.

Mapping by Design: A Guide to ArcGIS Maps for Adobe Creative Cloud

In *Three, Two, One. Climb On, Vancouver!* (figure 5.10), I show the climbing areas within three hours of Vancouver, British Columbia, using three buffers of travel time. Using the extension-direct Maps for Adobe Creative Cloud workflow, I created this monochromatic map by placing a single point in Vancouver, British Columbia, using the Add Places workflow. Then, I visualized the travel times in the same manner that you have just visualized travel times from the Winnipeg Airport, although in my *Three, Two, One. Climb On, Vancouver!* map, I chose one-, two-, and three-hour drive times.

Tutorial 5.2: Visualize routes to rock-climbing destinations

Scenario

You will be making a map of selected climbing destinations that are all located in the Lower 48 in the United States. The map will also include a small town in the state of Kansas that is not among the climbing destinations. You will be using the Visualize Route geo-analysis function to identify the driving routes from the town of Lebanon, Kansas, to each of the selected climbing destinations.

Learning objectives

- **Map setup review:** Create a mapboard for a specific project, and locate and add a web map hosted in ArcGIS Online using the extension-direct workflow

- **Advanced steps for map projection:** Gain a deeper understanding of the map projection settings in Maps for Adobe Creative Cloud

- **Adding a raster layer to reprojected mapboards:** Use raster data in reprojected maps

- **Geo-analysis—visualizing routes:** Visualize the travel routes between points of interest

Map setup review

Draw your mapboard around the contiguous United States

Before you draw the mapboard, you may want to use the Topographic basemap. It can help you see where to draw the mapboard.

1 Use the basemap selector to add the Topographic basemap.

2 On the Mapboards panel, draw your mapboard to fit this map's area of focus, which is the contiguous United States.

Make sure it covers the same or a similar extent as figure 5.11. And make sure that it includes the Lower 48.

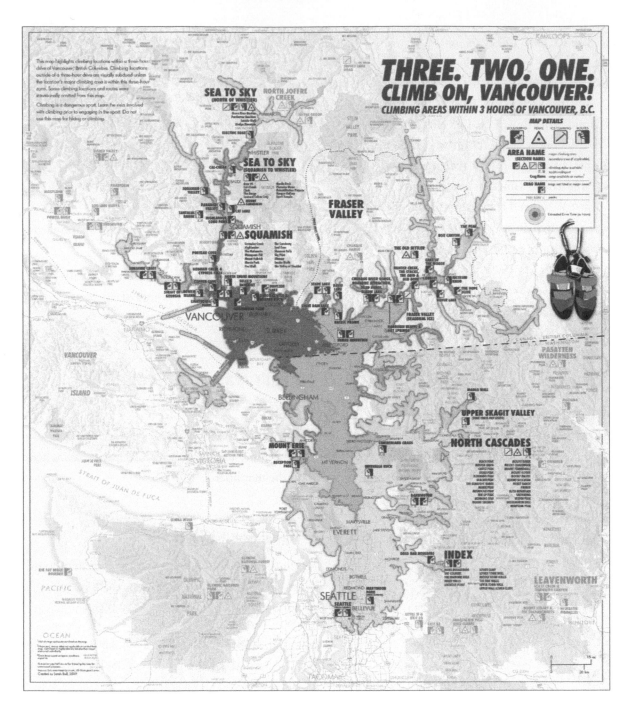

Figure 5.10. I created this map using Maps for Adobe Creative Cloud. The polygons radiating from Vancouver, British Columbia, show how far a person can drive from the heart of Vancouver within one, two, and three hours during normal drive times.

Mapping by Design: A Guide to ArcGIS Maps for Adobe Creative Cloud

Figure 5.11. The mapboard for this tutorial covers this same or similar extent of the Lower 48.

3 Name your mapboard **Visualize-Routes-to-Rock-Climbing,** and set the level of detail to **1:12,000,000.**

Note that your artboard size and units may vary slightly from the Mapboard Options dialog box in figure 5.12.

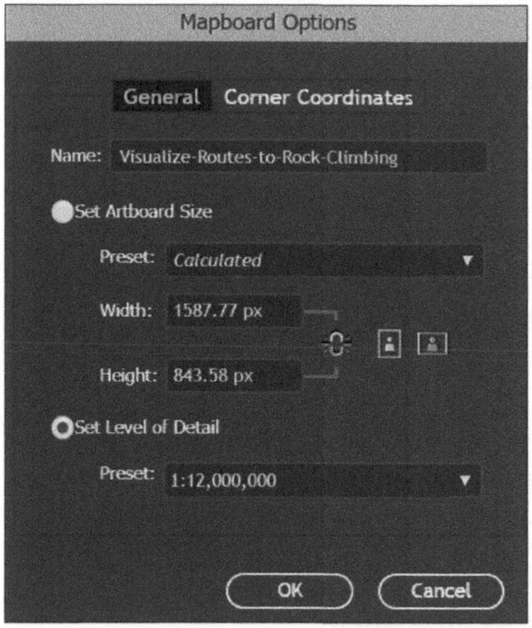

Figure 5.12. The Mapboard Options settings for the Visualize-Routes-to-Rock-Climbing map.

Add data to the Visualize-Routes-to-Rock-Climbing mapboard

1 Open the Compilation panel. After doing so, if you cannot see the full mapboard's extent in the map preview area, press Ctrl + 0 (PC) or Cmd + 0 (Mac) to fit the map to the preview area.

2 On the Compilation panel, click Add Content, and then click Overwrite From Map since you will be adding a web map to the compilation.

Make sure you choose this overwrite option and not the Add Layers option.

3 In the Overwrite From Map window, select the ArcGIS Online library, and then in the Search text field, type **Tutorial_5_2_VisualizeRoutes**. Press Enter or Return to yield the search results.

Of these results, you will be adding the item that has the thumbnail image with "5.2 Climbing" shown in figure 5.13.

Figure 5.13. Thumbnail image for Visualize-Routes-to-Rock-Climbing map.

4 Click to add the web map to the Visualize-Routes-to-Rock-Climbing mapboard. You will receive a message letting you know the current map compilation will be overwritten. Click OK.

5 Click OK to accept the overwrite, and click Close to close the Overwrite panel.

After doing so, the map preview area should look like figure 5.14. The colors in your map preview area may vary.

Figure 5.14. The map preview area after the Tutorial_5_2_VisualizeRoutes web map is added.

Advanced steps for map projection

Reproject the map

For this map, you will be using an equal-area projection.

1 From the Compilation panel, click the Map Settings button to open the current Map Settings window.

2 Click the Projection tab.

3 In the search bar, type **USA Contiguous Albers**.

This filter will narrow the results to only the projections that contain that string of text.

4 Select USA Contiguous Albers Equal Area Conic from the list (figure 5.15), and click OK to reproject your map.

Allow a few moments for reprojection.

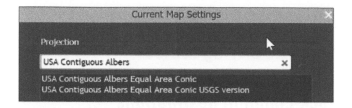

Figure 5.15. Choice for the USA Contiguous Albers Equal Area Conic projection.

If you did not remove the raster basemap before reprojecting, the raster basemap that was part of the web map will be removed during reprojection. The Action bar will alert, "Map layer(s) loaded with an error." Rather than an error, it is really an expectation; many raster layers' types are removed from the compilation during reprojection. In the next section, you will add a raster layer to your reprojected mapboard.

Adding a raster layer to reprojected mapboards

Search and add an image service

1 Click Add Content, and then select Add Layers. In the Add Layers window, select the ArcGIS Online library.

2 In the search bar, type **World Street Map**, and press Enter or Return.

3 Click the Filters button, and select Show Raster Layers.

4 Select the World Street Map raster map service by Esri (figure 5.16).

When searching, make sure that the results are sorted by highest rated. You may need to scroll to locate the layer in the results.

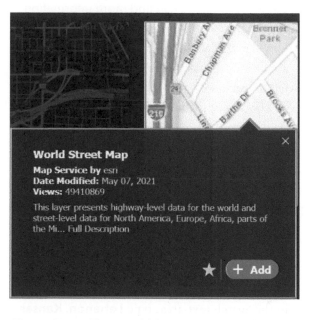

Figure 5.16. The World Street Map thumbnail provides information about the web map.

5 Add the World Street Map layer to your map.

6 Close the Add Layers panel.

Map data information

Did you know that you can find more information about a layer or web map by clicking the item's thumbnail in the Add Layers window results list? Metadata such as the date the layer was last modified and the item description can help you identify geographic data that are useful to your map's purpose. Click Full Description to learn more about the data layer or web map.

Visualize driving routes from Lebanon, Kansas, to the rock-climbing destinations

In this section, you will identify driving routes from Lebanon, Kansas, to each of the rock-climbing destinations on your map.

1 On the Contents panel, select Visualize Route from the Lebanon, Kansas, USA layer options.

2 On the Visualize Route dialog box, make sure that the Rock_Climbing_Destinations layer is selected in the first option, Route To Destinations In (figure 5.17).

3 Select Driving Distance as the measure.

4 For the route shape option, select Follow Streets.

Add the city of Lebanon, Kansas, to your map

1 In the Add Content options, select Add Places.

2 In the Search text area, type **Lebanon, Kansas,** and press Enter or Return.

3 Click the Add button for Lebanon, Kansas, to add a point to your map for that location.

You will need to close the Add Places dialog box to see the point on your map.

This designation is important for your results to follow driving routes. If you selected Straight Line, the resulting layer indicating the connection from Lebanon to the climbing destinations would be straight lines and not necessarily drivable.

5 In Result layer name, rename the layer **Driving routes.**

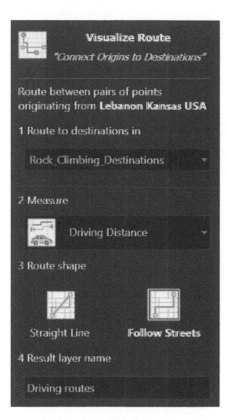

Figure 5.17. The Visualize Route settings to connect driving routes between Lebanon, Kansas, and rock-climbing destinations.

 Click Apply to perform the geo-analysis.

This will take a few moments. Please be patient.

 Once the routes are identified, click OK to add these new routes to your map.

After following the steps in the "Geo-analysis—visualizing routes" section, your map should look like figure 5.18. The color of the routes and point data layers may vary.

Finalize your map and sync it

Take a moment to consider what other information you may want on this map. Do you want to add labels for the states and provinces? Do you want to keep the current basemap? You can even add labels for the driving routes layer that you created in the preceding "Visualize driving routes from Lebanon, Kansas, to the rock-climbing destinations" section. Or perhaps the current map is set up exactly how you want it. Using the Maps for Adobe Creative Cloud skills that you have acquired, along with your personal vision for this map, make any changes and updates you want, and then sync the Visualize-Routes-to-Rock-Climbing mapboard to create an AI file. Once you have synced your map, on your Visualize-Routes-to-Rock-Climbing.ai layers panel, you can find your driving routes as separate sublayers within your Driving routes layer (figure 5.19). The order of your routes may vary.

The *Climbing Destinations* map (figure 5.20) I created from the "Geo-analysis—visualizing routes" section shows routes from the geographic center of the country, Lebanon, Kansas, to climbing destinations across the country.

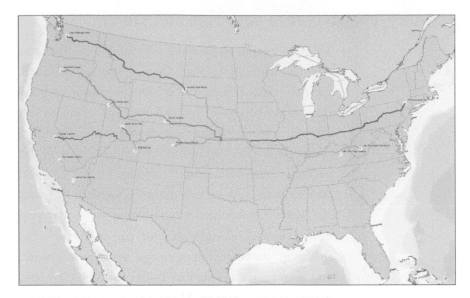

Figure 5.18. The Visualize-Routes-to-Rock-Climbing mapboard after using the Visualize Route function.

👁	⌄ ▢ Driving routes	◉
👁	› ▢ Route 1 - Lebanon, Kansas - The Shawangunks, New York	○
👁	› ▢ Route 2 - Lebanon, Kansas - Eldorado Canyon, Colorado	○
👁	› ▢ Route 3 - Lebanon, Kansas - Smith Rock, Oregon	○
👁	› ▢ Route 4 - Lebanon, Kansas - Wild Iris, Wyoming	○
👁	› ▢ Route 5 - Lebanon, Kansas - Joshua Tree, California	○
👁	› ▢ Route 6 - Lebanon, Kansas - The Needles, California	○
👁	› ▢ Route 7 - Lebanon, Kansas - Red River Gorge, Kentucky	○
👁	› ▢ Route 8 - Lebanon, Kansas - New River Gorge, West Virginia	○
👁	› ▢ Route 9 - Lebanon, Kansas - Southeast Utah	○
👁	› ▢ Route 10 - Lebanon, Kansas - Index, Washington State	○
👁	› ▢ Route 11 - Lebanon, Kansas - Yosemite, California	○
👁	› ▢ Route 12 - Lebanon, Kansas - Maple Canyon, Utah	○
👁	› ▢ Route 13 - Lebanon, Kansas - City of Rocks, Idaho	○
👁	› ▢ Route 14 - Lebanon, Kansas - Spearfish, South Dakota	○

Figure 5.19. The Visualize Route geo-analysis function creates a parent layer for the routes, and a sublayer is created for each route.

Mapping by Design: A Guide to ArcGIS Maps for Adobe Creative Cloud

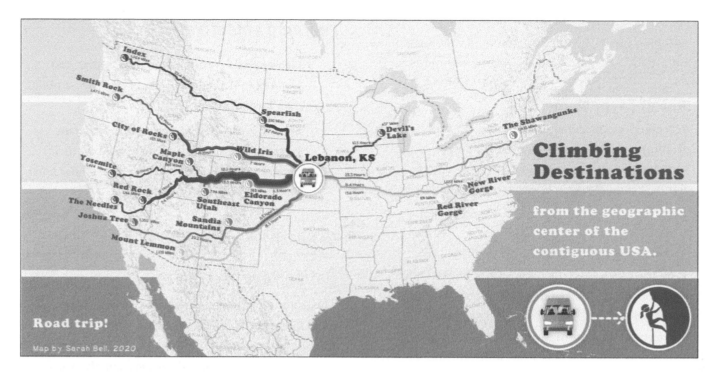

Figure 5.20. I created this map by following the "visualize routes" steps from tutorial 5.2, and then styled it in Illustrator.

MAPMAKER TIP

Add Demographic Data settings

When using the Add Demographic Data operation, you are creating a new layer enriched with demographic attributes. On the Add Demographic Data panel, there are four settings for the newly enriched layer: Select Variables, Define Areas To Enrich, Result Layer Name, and Define Default Labels. Some of these settings might be unavailable depending on the data type (point, line, or polygon) that you are enriching.

Select Variables

Select Variables opens the data browser, in which you can search for popular demographic variables to incorporate in the new resulting layer.

Define Areas To Enrich

When demographic information is added to point and line layers, buffers are calculated around the features. The Define Areas To Enrich setting defines the buffer parameters. The variable information is then retrieved for these buffer areas, and a new resulting point or line layer is created with this information as one of the new layer's attributes.

Result Layer Name

You can provide a custom name for the Add Demographic Data operation's resulting enriched layer. If you do not provide a custom name, a default name, Enriched [Layer Name], is provided.

Define Default Labels

With the Define Default Labels box checked, the first possible field from this operation's resulting enriched layer will be the default label, and it will appear on your newly created enriched layer. You may update the field from which the labels are drawn at any time, even after the enriched layer is created.

Tutorial 5.3: Add demographic data

Scenario

The final geo-analysis component that this chapter covers is Add Demographic Data. This feature uses ArcGIS GeoEnrichment[SM] Service, which applies demographic information to map layers. In this tutorial, you will be using the Add Demographic Data feature to enrich a point layer of cities in Australia as you create a thematic map.

Learning objectives

- **Geo-analysis—enriching data:** Use the Add Demographic Data operation to enrich a point data layer

- **Sorting layers by geometry type:** Sort data layers on the Contents panel based on their data type, in which points are above lines and lines are above polygons

Geo-analysis—enriching data

Define your mapboard

This tutorial requires using your Plus account, ArcGIS Online account (GIS Professional or Creator), or the free trial provided with this book. Enterprise users with the appropriate portal settings can also use these tools.

1. Open Illustrator, and sign in to Maps for Adobe Creative Cloud.

2. On the Mapboards panel, draw a mapboard around the country of Australia (figure 5.21).

3. Name your new mapboard **Australia,** and set the map scale to **1:15,000,000** (figure 5.22).

Your artboard size and units of measurement may vary, but that's okay.

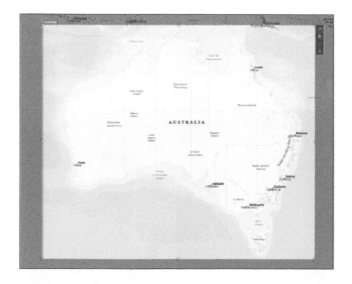

Figure 5.21. In this mapboard extent for Australia, the Topographic basemap is selected for its useful reference information.

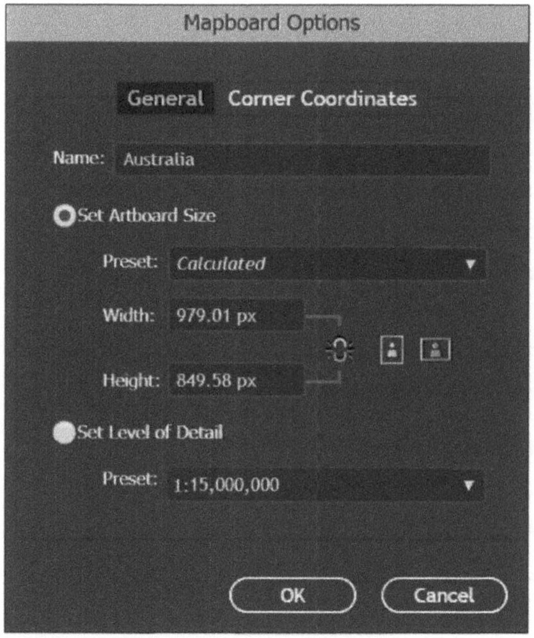

Mapboard Options

General   Corner Coordinates

Name: Australia

◉ Set Artboard Size

Preset: *Calculated*   ▾

Width: 979.01 px

Height: 849.58 px

◯ Set Level of Detail

Preset: 1:15,000,000   ▾

(OK)   (Cancel)

Figure 5.22. Mapboard options set for map scale.

Add map layers to the Australia mapboard

1 On the Compilation panel, add the World Countries (Generalized) layer. Click Add Content > Add Layers. From the Add layers panel, look in the library Living Atlas of the World > Boundaries > Administrative, and search for **World Countries (Generalized)** (figure 5.23).

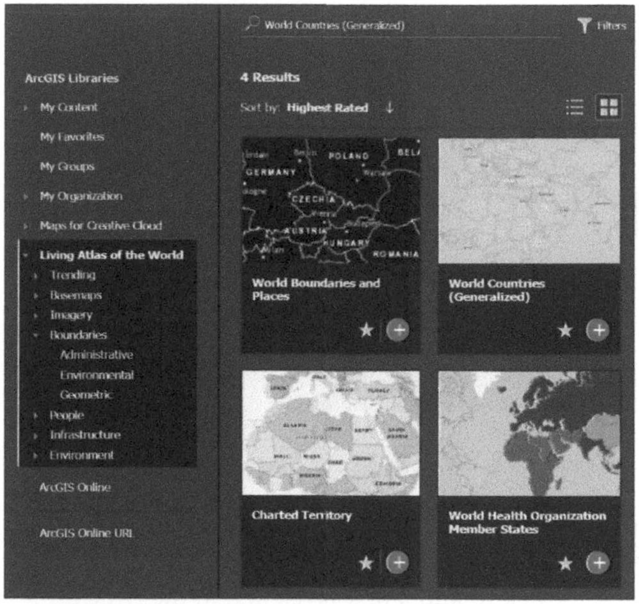

Figure 5.23. From the Living Atlas of the World library, the World Countries (Generalized) layer is added to the map. To narrow the results, you can search from the Administrative group of the ArcGIS Living Atlas's Boundaries subcategory.

2 Add the World Administrative Divisions layer to the Australia mapboard.

This polygon layer contains high-level regions, such as provinces and states. You can narrow your results for this layer by searching from the same Living Atlas subcategory that you searched from under Boundaries > Administrative when you added the World Countries (Generalized) layer. The item thumbnail for the World Administrative Divisions layer is shown in figure 5.24.

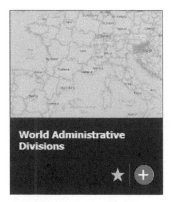

Figure 5.24. The World Administrative Divisions layer from the Living Atlas library.

3 Click to add a new layer, and in the ArcGIS Online library search area, type **Australia_Cities_MappingByDesign** and click the Search button.

This layer (figure 5.25) is a point dataset of cities in Australia.

4 On the Contents panel, double-click the layer name, and rename the Australia_Cities_MappingByDesign layer **Australia Cities**.

Figure 5.25. Result from a search for Australia_Cities_MappingByDesign from the ArcGIS Online library.

Create a filter to show only Australia's administrative boundaries

Because the focus of this map is the country of Australia, you will be creating a filter that removes the administrative boundaries for all other countries.

1 On the Contents panel, select Filter from the World Administrative Divisions layer options.

2 On the Filter panel, use the drop-down list, and type to set the expression to **Country is Australia**. Click Add Another Expression, and add another expression that reads **Land_Rank is 5**.

③ Once your filter dialog box appears as it does in figure 5.26 with both expressions set, click Apply Filter.

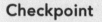

Figure 5.26. The filter in this dialog box will display only Australia's states and will hide all other countries' administrative divisions.

④ Rename the newly filtered World Administrative Divisions layer **Australia States**.

Checkpoint

At this point in the tutorial, the layers on your Contents panel should be as they appear in figure 5.27. Your basemap may vary from the one depicted in this image. That is fine.

Figure 5.27. The Contents panel should appear with four layers at this point in the tutorial.

Enrich the Australia States polygon layer with income data

① From the Australia States layer options, choose Add Demographic Data.

Choosing this layer will open the Add Demographic Data panel (figure 5.28).

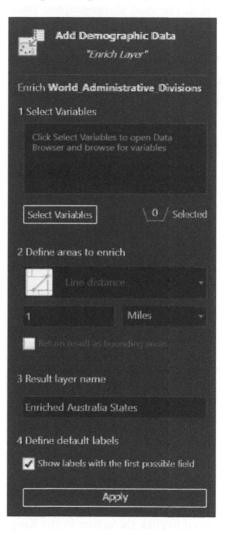

Figure 5.28. The Add Demographic Data panel in Maps for Adobe Creative Cloud.

② Open the data browser by clicking Select Variables.

③ In the data browser, select Australia from the location drop-down list, and select the Income theme (figure 5.29).

This theme will return popular variable results that can be incorporated in the Add Demographic Data operation.

Figure 5.29. The Add Demographics Data browser with Australia selected as the region.

④ Browse more income-related variables by clicking Purchasing Power (figure 5.30).

Note that you may need to click the arrow in the Keep Browsing section to see the Purchasing Power option.

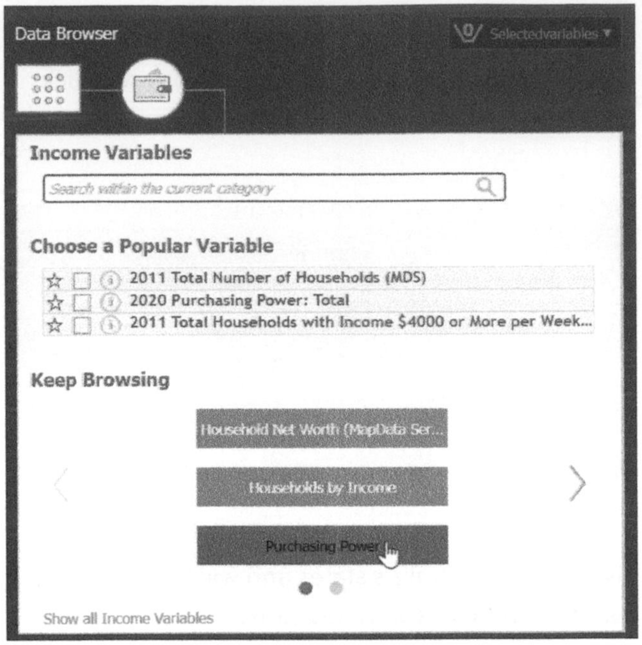

Figure 5.30. Browse even more income-related variables by clicking Purchasing Power.

⑤ In the resulting Purchasing Power variables, check the box for 2020 Purchasing Power: Per Capita (figure 5.31), and then click Apply in the data browser.

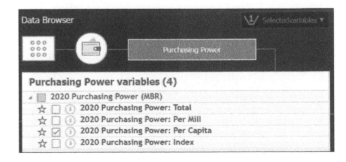

Figure 5.31. Box checked for 2020 Purchasing Power: Per Capita.

6 In Result Layer name, type **Purchasing power Australia states**, and then click Apply to create the newly enriched layer.

Creating the layer (figure 5.32) will take a moment. Once the analysis is complete, you will receive an alert.

Figure 5.32. The Add Demographic Data panel with Purchasing power Australia states set for the layer name for the layer resulting from this geo-analysis.

7 After the analysis is complete, click OK on the Add Demographic Data panel to add the new Purchasing power Australia states layer to your map. You may need to scroll to the bottom of the panel to see the OK button.

Sorting layers by geometry type

Reorder layers

You now should have a new polygon layer at the top of your contents called Purchasing power Australia states. Because this new polygon layer is above the point data layer, you will use the Layer Reorder button to quickly sort the layers by geometry type.

1 Click the Layer Reorder button once to sort layers by geometry type.

The Layer Reorder button is located at the bottom of the Contents panel next to the Delete button.

Because you have just reordered the layers by geometry type, the layers are logically stacked, and so the Layer Reorder button will become unavailable.

Reproject the Australia mapboard

1 Delete the basemap.

This thematic map does not require a basemap. Delete the basemap from the Contents panel before reprojecting the map.

2 Click the Current Map Settings button to open the dialog box.

3 Click the Projection tab.

4 In the search area, type **Australia,** and select the GDA 1994 Australia Albers projection from the results (figure 5.33). Click OK.

5 Once the map is updated, show the entire view in the map preview area by selecting Fit On Screen from the Quick Zoom drop-down list.

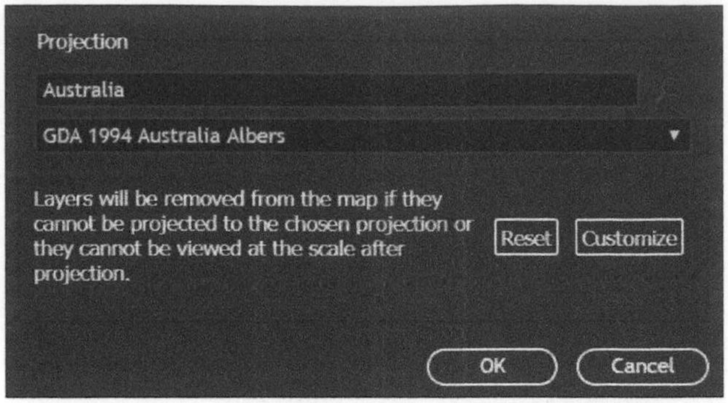

Projection

Australia

GDA 1994 Australia Albers ▼

Layers will be removed from the map if they cannot be projected to the chosen projection or they cannot be viewed at the scale after projection.

Reset Customize

OK Cancel

Figure 5.33. The Australia map in this tutorial will use the GDA 1994 Australia Albers projection.

Checkpoint

At this point in the tutorial, your Contents panel should have the same layers as listed in figure 5.34—that is, Australia_Cities_MappingByDesign – Australia_Cities, Purchasing power Australia states, Australia States, and World Countries (Generalized) – World_Countries_(Generalized). The map should look roughly as it appears in this figure as well, although the colors and extent may vary. Make sure that the entire country of Australia is within the map extent.

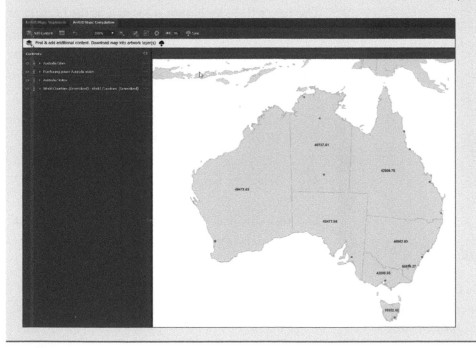

Figure 5.34. The status of the Compilation panel after completing reprojection.

Turn off syncing for the Australia States layer

You have created a filter for the original World Administrative Divisions layer to show only the Australia states, and then you renamed the layer Australia States. From this Australia States layer, you created a new polygon layer of Australia states to which you added purchasing power data. Australia will be symbolized by this new enriched layer; therefore, the original Australia States layer is no longer necessary for your map.

 From the Australia States Layer Options menu, click the down arrow to the right of the layer name to toggle off syncing for the Australia States layer (figure 5.35).

Figure 5.35. The down arrow next to Australia States with syncing toggled off for the Australia States layer.

Finalize and sync your map

You are now ready to add final details to your map, and then sync it. What are some elements or features that you would use to enhance this map? Now that the mapboard is set to the final map projection, perhaps you want to add a new basemap or other additional data layers. Would labels help orient your map readers?

 Using what you have learned thus far, enhance your Australia map in a way that makes sense for potential map readers.

2 Once you have added your desired finishing touches, sync the mapboard to create a new Australia.ai file.

> ### Reminder
> You can also use the Processes panel to enhance the Australia.ai file during and after syncing.

Shaded relief

The varying topography of Earth's surface can be captured by maps in several ways, shaded relief being among the most common. Shaded relief, also known as hillshade, renders terrain as if it were illuminated by a light source. This mimicked light source is usually depicted as shining from the upper left of the map and toward the lower right. The reason that a shaded relief's light source is almost always represented as illuminating the landscape from the upper part of the map is that we are accustomed to seeing our world lighted from above. With shaded relief rendered in this way, our brains correctly interpret the terrain. When a shaded relief's light source is depicted from the bottom part of the map, it can cause human misinterpretation of the landscape, whereby our brains read the landscape as inverted from reality—the low parts look high, and the high parts look low. This optical illusion is known as relief inversion.

Because a map typically shows north at the top and because shaded relief is usually illuminated from the upper (most often upper left) part, shaded relief is typically illuminated from the northwest. However, even when north is not at the top of a map, shaded relief illumination is still typically from the upper part of the map. Figure 5.36 illustrates this phenomenon of relief inversion.

In real life, when we are viewing actual terrain during the day, there is one main light source—the sun. However, the sun's light scatters by reflecting off illuminated slopes and bouncing onto other slopes. To capture this scattered light, some mapmakers prefer using a multidirectional hillshade as opposed to hillshade rendered using a single light source. These multidirectional raster layers can be created computationally in a GIS application such as ArcGIS Pro. ArcGIS Online offers users a multidirectional hillshade layer that mimics six different light sources, which prevents overexposing northwestern-facing slopes or underilluminating southeastern-facing slopes. The Kananaskis overview map by contest winner Ian Ladd (figure 5.37) was created using Maps for Adobe Creative Cloud, in which Ladd advantageously used the multidirectional hillshade layer available in ArcGIS Online.

Figure 5.36. These two identical shaded-relief images of Mount St. Helens differ only in their rotation. In the right version, the top of the image is north, and the light source is at the upper left. This is the most common azimuth for illuminating shaded-relief terrain. On the left, the image has been rotated 180 degrees so that south is at the top of the image, and the light source is at the lower right. This orientation gives the image an inverted look, as if the mountains sink into the earth like canyons.

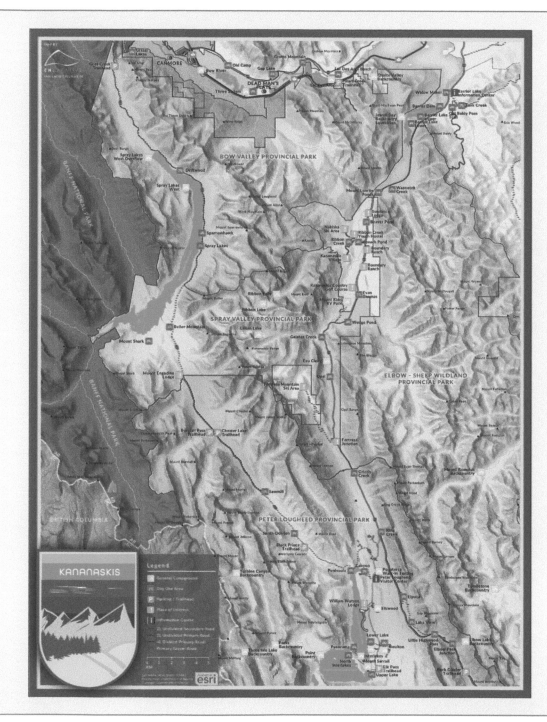

Figure 5.37. The Kananaskis overview map by Calgary-based cartographer Ian Ladd is a wonderfully designed representation of this wild area in Alberta, Canada. Created using Maps for Adobe Creative Cloud, this overview map enhanced by hillshade is inspired by Ladd's passion for mapmaking and outdoor recreation.

CHAPTER 6

CUSTOMIZING MAPS USING THE PROCESSES PANEL

B Y THE TIME YOU REACH THE END OF THIS chapter, you will have been introduced to, and even mastered, many common workflows for building maps directly in Maps for Adobe Creative Cloud. The next and final chapter in this book is dedicated to the ArcGIS Pro–to–Illustrator workflow. Although chapter 7 covers even more Maps for Adobe Creative Cloud map-creation features, rather than starting the map from Maps for Adobe Creative Cloud, the initial map creation in the final chapter begins in ArcGIS Pro, Esri's desktop GIS application.

Chapter 6 is entirely within Maps for Adobe Creative Cloud, with a thorough investigation of the Processes panel tools. In fact, rather than calling them tools for a moment, let's regard these beneficial functions for what they are: processes, hence the name of their panel. As you will soon discover, the processes from the Processes panel that are covered in this chapter are robust, time-saving map design operations. Some of this chapter's single tutorial will be a light review. In chapter 3, you created a map legend from the Processes panel, and in chapter 4 you used the custom symbols replacement process. In this tutorial, you will review those processes and explore new ones.

Tutorial 6.1: Mapping the Olympic Peninsula

Download the essential Illustrator libraries before you begin

This tutorial requires three Adobe Illustrator libraries, which must be downloaded from the web. The items listed in the following instructions are zipped folders that contain a single AI library. Although their item type is listed as Image Collection, these items are AI .zip files.

The exercise data for this chapter (figure 6.1) are available at links.esri.com/MappingByDesignData. They are shared with the ArcGIS Online group Mapping By Design (Esri Press) in the Learn ArcGIS organization.

1 Download the following data:

- Tutorial_6_1_SymbolsLibrary.zip
- Tutorial_6_1_BrushesLibrary.zip
- Tutorial_6_1_SwatchLibary.zip

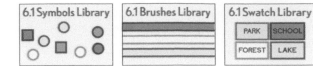

Figure 6.1. The thumbnail images for the Illustrator Symbols, Brushes, and Swatch Libraries hosted in ArcGIS Online.

(2) Once you have downloaded the zipped folders containing Illustrator libraries, unzip each one, and save the AI files to a location that you will remember, such as EsriPress\MappingByDesign.

You will be accessing these files during the following tutorial.

Scenario

In tutorial 6.1, you will use each of the symbology processes—symbols, brushes, and swatches—as you build a map of Washington State's Olympic Peninsula in the Pacific Northwest region of the United States. You will also learn how to concatenate large batches of unconnected line segments into single paths using the join lines process.

One more thing, to alleviate any confusion, note that although the Processes panel functions are, in fact, processes, this book sometimes refers to them as tools. Now let's use these tools to make a map.

Learning objectives

- **Web maps review:** Create an initial mapboard from a web map's extent

- **Corner coordinates:** Refine mapboard extents with precise geographic coordinates using Corner Coordinates mapboard settings

- **In-sync custom symbology processes:** Use custom symbology replacement scripts during a map sync

- **Brushes process:** Replace map line work with styles from an Illustrator brushes library

- **Join lines process:** Use the join lines process to concatenate large batches of an Illustrator layer's unconnected lines

- **Swatches process:** Replace map polygons with styles from an Illustrator swatches library

Part 1, Map setup and sync

In this first part of tutorial 6.1, after reviewing how to create a mapboard from a web map, you will further refine your mapboard extent with precision as you manually enter latitude and longitude coordinates on the mapboard settings' Corner Coordinates tab. You will also use the brushes process to automatically apply Illustrator library artwork to a linear data layer as the mapboard is synced.

You are becoming proficient in Maps for Adobe Creative Cloud workflows. This tutorial navigates the Processes panel; however, the rest of the map creation is up to you. You will be guided through much of the mapmaking workflow, but you will not be instructed to add labels or additional layers. Of course, if you feel that the map needs labels—or any other enhancements not included in the tutorial—feel free to include them. That is, this tutorial can purely be an exercise for learning the Processes panel, or you can use it as an opportunity to design a complete, beautiful map of the Olympic Peninsula.

Web maps review

Create a mapboard from a web map

 On the Mapboards panel, click Import, and select From WebMap.

 On the Create From Map dialog box, click ArcGIS Online to search from that library, and in the Search text area, type **Tutorial_6_1_ Olympic_Peninsula** (figure 6.2). Then press Enter or Return.

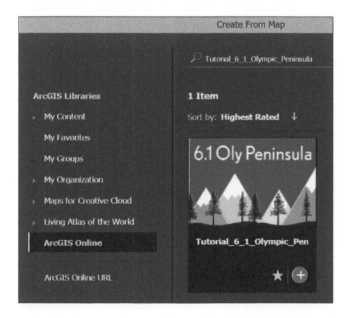

Figure 6.2. The search criteria return this web map containing map layers for the Olympic Peninsula in the Pacific Northwest region of the United States.

 Click the Add button on the web map item.

 Once the web map is loaded, close the Create From Map dialog box.

Corner coordinates

Update the mapboard extent using the Corner Coordinates option

ArcGIS web maps are essentially a comprehensive collection of map layers hosted in ArcGIS Online. When a web map is saved to ArcGIS Online, the web map's extent is also saved. In Maps for Adobe Creative Cloud, whenever a mapboard is created from a web map, the extent that is saved to the web map determines the mapboard's extent. In tutorial 4.1, you discovered that this extent can be manually resized by dragging the mapboard edges' control points. Here, you will be adjusting the extent by assigning latitude and longitude coordinates for the mapboard corners.

 On the vertical mapboard toolbar, click the Mapboard Options button (second from the top) to open the Mapboard Options dialog box.

 Rename the mapboard **Olympic_Peninsula** so that your AI file will be named Olympic_ Peninsula.ai.

 Click the Corner Coordinates tab to show the coordinate selector.

You can view the latitude and longitude coordinates of your map extent's corners by clicking any corner of the coordinate selector. The coordinates are expressed in decimal degrees and can be edited by typing a new value in the text areas. If an invalid value is entered, you will receive an alert requesting a new value.

4 Click the upper-left corner in the coordinate selector.

This corner corresponds to the map's northwest coordinates.

As discussed in chapter 5, Maps for Adobe Creative Cloud uses Web Mercator as the default projected coordinate system. As such, across the entire map, north is up, or at the top of the map. This means that if you drew a line from any point on the map to north, it would be a straight line perpendicular to the equator. Therefore, by editing the longitude of the northwest corner, you are also editing the longitude of the lower-left (southwest) corner because the northwest and southwest coordinates will always share the same longitude on this map. Likewise, editing the latitude of this northwest coordinate will change the latitude of the northeast coordinate, as their latitudes are identical.

5 Type **48.48** to change the northwest corner's latitude and **−124.87** to change the longitude (figure 6.3).

Do not click OK yet, as this will close the window.

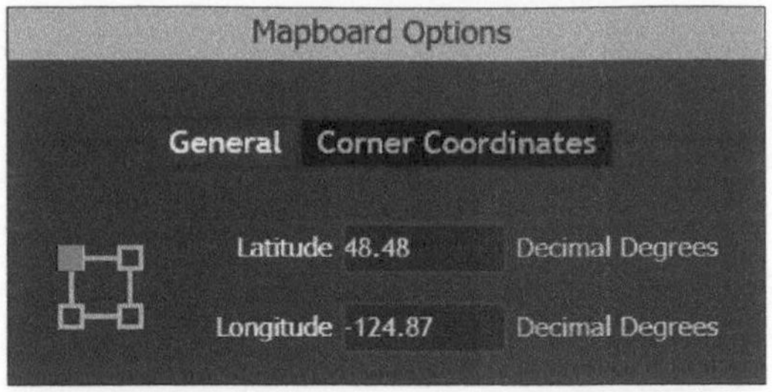

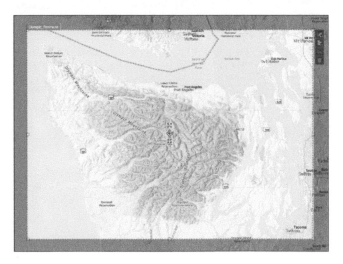

Figure 6.3. The northwest coordinates for tutorial 6.1's mapboard.

6 Select the lower-right (southeast) corner on the coordinate selector, and update the latitude to **47.20** and the longitude to **−122.25**.

7 Click OK on the Mapboard Options dialog box to apply your edits to the Olympic_Peninsula mapboard coordinates.

After updating the mapboard's corner coordinates, your mapboard's extent should match the extent in figure 6.4, although your basemap may differ.

Figure 6.4. The mapboard extent after updating the corner coordinates.

Mapping by Design: A Guide to ArcGIS Maps for Adobe Creative Cloud

Delete the raster basemap

1 Open the Compilation panel, and delete the raster basemap from the Contents panel.

On the Compilation panel's map preview area, the large bodies of water may appear as black polygons. This is because of the extreme contrast between the very bright polygons in the United States State Boundaries 2018 layer and the Imagery raster basemap that was in the web map you added. By turning off the visibility of the state boundaries layer, the imagery of the basemap becomes more apparent. For this step, you are removing this raster basemap just as you did in previous tutorials in the book.

On the Compilation panel toolbar, update the PPI

1 Click in the PPI text area to make it editable, and type **200**.

Set up the line layer symbology for custom artwork during syncing

1 On the Compilation panel, rename the Olympic Roads layer **Roads** (case sensitive).

Roads is one of the brush names from the Tutorial_6.1_BrushesLibrary.ai library that you were instructed to download at the beginning of this tutorial. For Maps for Adobe Creative Cloud to apply the correct brush from this library to a specified map layer, the map layer containing the soon-to-be-restyled features must have the identical name as the brush.

2 Rename the Olympic_Rivers_Streams line layer **Rivers** (also case sensitive).

Make sure that you are renaming the correct layer because two layers represent rivers in this map, one line layer and one polygon layer.

Once you have renamed the layers to match the brushes, the next step is to point to the AI file that contains the brush artwork.

3 On the Compilation panel toolbar, click the Processes button to open the Processes panel.

4 On the Processes panel, click Brushes. Then click Set Path, and navigate to EsriPress\MappingByDesign or where you saved the Tutorial_6.1_BrushesLibrary.ai file. Select the file to populate the Set Path text area.

5 Check the box next to Run On Sync so that it is on.

This will ensure that the rivers and roads symbology is replaced with the brushes during the map sync.

6   Back on the Compilation panel, click the
Sync button to sync the Olympic_Peninsula
mapboard, which will create an AI file of the
same name.

Be patient as your mapboard is loaded.

Checkpoint

Congratulations! You have reached the end of
tutorial 6.1, part 1. You created an AI file named
Olympic_Peninsula.ai, which contains layers that
were generated from the Tutorial_6_1_Olympic_
Peninsula web map. In this new AI file, the artwork
for the Roads and Rivers layers will be styled with
the brushes of the same name in the Tutorial_6.1_
BrushesLibrary.ai file. The AI file artboard should
appear as it does in figure 6.5.

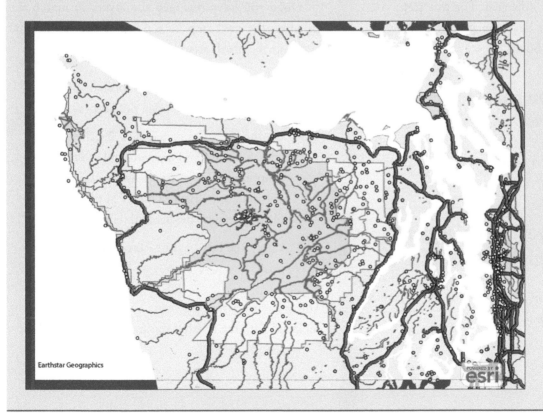

Figure 6.5.
Artboard from
downloading a new
Olympic_Peninsula.ai
file after completing
part 1 of tutorial 6.1.

Earthstar Geographics

Part 2, Using processes on a synced AI file

In part 2, you will be working with your downloaded Olympic_Peninsula.ai file. This section begins with an examination of the brushes applied during the map-board's first sync. Then you will use the join lines process to clean up some of the line work in your map, as you connect disjointed segments into easy-to-work-with paths.

Associated brushes added to AI file

Both brushes that you applied during the map sync have been added to the Olympic_Peninsula.ai file's Brushes panel. In this first section of part 2, you will examine these two brushes on your Brushes panel.

Brushes process

Explore the new brush styles in your Olympic_Peninsula.ai file

1. With Olympic_Peninsula.ai selected as the active file, open the Illustrator Brushes panel (Window > Brushes, or F5).

For more about keyboard shortcuts, see table 1 in the introduction to this book.

On the panel, you will find three brushes (figure 6.6): the Rivers and Roads brushes that you applied during the sync as well as a brush style named [Basic].

Figure 6.6. The Illustrator Brushes panel shows the brushes that were added to your map during the sync.

MAPMAKER TIP

Tools and shortcuts

Illustrator provides many user-friendly keyboard shortcuts that help make the design process efficient. For many of the tools, when you hover the pointer over its icon, a ToolTip will appear showing its keyboard shortcut.

2. With Illustrator's Selection tool , select one of the highway segments by clicking it.

The Illustrator keyboard shortcut for the Selection tool is V.

3. Open the Illustrator Color panel (Window > Color, or F6).

In Illustrator, when a brush is applied to a path's stroke, the basic style of the path's fill and color will not be lost; adding a brush is similar to replacing the actual path's style with a decorative mask. You can always remove the brush and return to the basic style. In step 4, you will inspect this basic style.

4. Open the color panel (F6). With the Roads segment still selected, check the color panel for this layer's stroke color by clicking the stroke graphic.

Notice that the color of the selected Roads path (figure 6.7) is set to 77, 77, 77 (RGB) or #4d4d4d (hexadecimal), and there is no fill color; this color scheme matches the original Roads layer settings from the Maps for Adobe Creative Cloud Compilation panel.

Because the Roads brush has been applied to this layer, the artwork on the map will match the brush. You can return the layer's artwork to its original simple style by clicking the [Basic] layer on the Brushes panel.

Figure 6.7. The stroke selected and the hexadecimal color value #4d4d4d.

5. With that Roads segment still selected, click the [Basic] item in the Brushes library.

This action will remove the brush from the selected highway segment.

The Roads layer original style still exists. This backup is useful for those moments when you decide that the brush style isn't right for your map.

6. Press Ctrl + Z (PC) or Cmd + Z (Mac) to undo, returning the selected segment to the brush appearance.

You can also try returning to the original color with the Rivers layer. By selecting one of that layer's paths on the map, and then clicking the [Basic] style on the Brushes panel, the artwork will return to its original blue color, #002673.

Join lines process

Concatenate segments using the join lines process

In chapter 3, you were introduced to the different geographic data types: points, lines, and polygons. In GIS applications such as ArcGIS Pro, how or if a geo-analytical operation will perform depends on the geographic data type to which it is applied. You encountered some of this dependency in chapter 5 when you created driving routes that connected cities from a point data layer; this operation will work only on point data.

In a vector graphic editing program such as Illustrator, there are no points, lines, or polygons, at least not in the GIS sense of these terms. In Illustrator, what we consider to be a geographic point may be represented by a small closed path, such as a circle. A polygon could appear as a geometric closed path, such as the artwork in the Olympic_Peninsula.ai file's Olympic_FederalLand layer. Geographic lines, such as the Olympic_Trails layer artwork, are typically unclosed paths in Illustrator. Often, in an AI file, the open paths that represent linear map features are disjointed.

When a map's aesthetic is designed in Illustrator, the disjointedness is not always necessary and can even hinder the appearance. In this section of part 2, you will be using the join lines process to connect the disjointed paths in the Olympic_Trails layer for a better design experience.

1. In Illustrator, make only the Olympic_Trails layer's artwork visible and hide the remaining layers' artwork. To do this, hold the Alt key and click the Olympic_Trails layer's visibility button.

If you already turned off a layer's visibility, you may need to click twice. Make sure the Olympic_Trails layer is unlocked.

2. Select all the artwork inside the Olympic_Trails layer by clicking the space to the right of the target icon for that layer.

In figure 6.8, the space is marked by a colored box on the Olympic_Trails layer on the Illustrator layers panel. Once you click in this area, a small square will appear indicating that artwork in the layer is selected. Notice that the color of your small square may differ from the figure. Also, notice that the Olympic_Trails parent layer, Olympic_Peninsula_Sync_1, has a colored box next to its name as well. This marking indicates that the artwork you selected is in this parent layer.

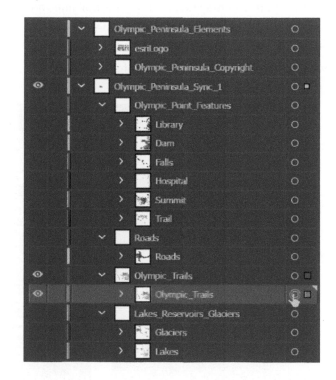

Figure 6.8. The colored square indicates that there is artwork selected in the Olympic_Trails layer. Clicking the square will select all unlocked artwork in this layer.

Smoothing lines

There are several ways to smooth the appearance of jagged-looking artwork in Illustrator. In fact, there is even an Illustrator tool called the Smooth tool. This map tip focuses on the Stroke panel to update your line features for a nice visual flow.

After completing the join lines process, make sure that the original Olympic_Trails layer's visibility is turned off. You may notice that your trail lines are jagged. The Olympic mountains are steep, after all. For those of you who have hiked steep mountains, you know that there will be a lot of switchbacks.

For your map, this means that trail lines will have a lot of small, sharp turns that represent these switchbacks. The strokes in Illustrator typically default to a miter join, which gives these sharp turns a pointy appearance. You can tone down the pointiness by updating this setting on the Stroke panel (figure 6.9). With the artwork in the Olympic_Trails layer selected, open the Stroke panel (Window > Stroke or Ctrl + F10 [PC] or Cmd + F10 [Mac]). Click the Round Join button as the new corner setting.

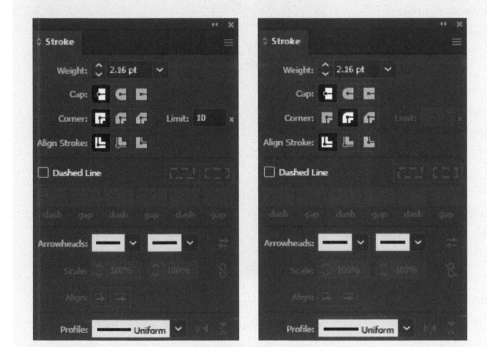

Figure 6.9. *Left*, the corner is set as a miter join, whereas, *right*, the corner is set as a round join.

3 With all the artwork in only the Olympic_Trails layer selected, open the Maps for Adobe Creative Cloud Processes panel and select Join Lines.

Selecting it will not run the process.

4 With Join Lines selected, click the Join Lines button at the lower right of the panel to begin the process.

This may take a few moments depending on your computer's processing power.

After the Join Lines operation is completed, you will have a new layer named Olympic_Trails Join, which contains the newly concatenated trail lines. The original Olympic_Trails layer remains and is located beneath the new layer.

Replace the default point symbols with the custom point symbols post sync

1 Isolate the layer visibility of each of the sublayers of the Olympic_Point_Features parent layer, so that the artwork in only those sublayers is visible (figure 6.10).

Figure 6.10. Layer visibility isolated for Olympic_Point_Features sublayers.

2 In Maps for Adobe Creative Cloud, open the Processes panel, and click Symbols to activate that process.

Clicking Symbols will not run the process.

3 Set the path to the Tutorial_6.1_SymbolsLibrary.ai file that you saved to your computer.

Make sure that Run On Sync is checked off since you are running this on an already-synced map.

The symbol names in the Illustrator library already match their respective Illustrator layer names. For this reason, you do not need to update the Illustrator layer names.

4 Click Replace Symbols to initiate the symbol replacement.

Replacing symbols may take a moment.

As you discovered in the first section of part 2, exploring new brush styles, applying a style from an Illustrator library will automatically add the style to the respective Style panel for the AI file that you are working in. Each of the symbols that you applied to the Olympic_Point_Features sublayers has been added to the Olympic_Peninsula.ai Symbols panel. Open the Symbols panel (Window > Symbols) where you will find each of the symbols (figure 6.11).

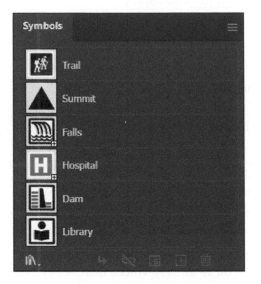

Figure 6.11. The Olympic_Peninsula.ai file's Symbols panel contains each point symbol added to the map when the default point symbols were replaced with custom point symbols.

Replace the default polygon artwork using custom swatches

1. Turn on the visibility for all artwork on the Layers panel. You can do this by holding the Alt key and clicking any layer's visibility button.

2. Rename the following Olympic_FederalLand sublayers as follows:

   - U.S. Forest Service to **Forest Fill**
   - National Park Service to **Park Fill**

3. Rename the Lakes sublayer (located in the Lakes_Reservoirs_Glaciers parent layer) **Water Fill**.

4. In Maps for Adobe Creative Cloud, click Swatches from the Processes panel to activate this process.

5. Set the path so that it points to the Tutorial_6.1_BrushesLibrary.ai that you saved to your computer.

6. Click Replace Swatches on the lower left of the Processes panel to update your map symbology with the custom swatches.

A hack for making oceans

If you are making a map in Illustrator that has a lot of shoreline, such as the map you made in this chapter's tutorial, you will need a water body to fill the void. The Olympic_Peninsula.ai map has a lot of shoreline but no ocean to fill the empty space next to the shoreline. You can create an ocean with the following simple trick:

1. Create a new layer in Illustrator and drag it to the bottom of the stack so that it sits beneath the United States layer.

2. Rename it **Ocean**.

3. Select the layer name to highlight it.

4. This will activate the Ocean layer.

5. In the Ocean layer, use the Rectangle tool to draw a rectangle that spans the entire artboard.

6. Give the rectangle a nice ocean-blue color, and voilà, you now have an ocean.

CHAPTER 7

THE ArcGIS PRO-TO-ILLUSTRATOR WORKFLOW

THIS CHAPTER FOCUSES ENTIRELY ON THE ArcGIS Pro–to–Illustrator workflow using Maps for Adobe Creative Cloud to open ArcGIS Pro–generated Adobe Illustrator Exchange (AIX) files as well-organized AI files. Because the Maps for Adobe Creative Cloud user base is broad—ranging from professional graphic designers to cartographers to GIS analysts—some of you reading this book may rarely use desktop GIS software, whereas others may use GIS in daily workflows. Even if you have never used ArcGIS Pro, this chapter offers insights into the collaborative association that Maps for Adobe Creative Cloud offers between GIS analysis and map design.

The integrated ArcGIS Pro–to–Illustrator workflow using AIX files can occur between a GIS analyst developing a basic map and passing it on to a designer, or it can be performed by a single mapmaker executing the GIS analysis, data management, and aesthetic map design. Projects developed in ArcGIS Pro (Esri's professional GIS desktop application) can be exported to an AIX file that Maps for Adobe Creative Cloud processes into a well-organized ready-to-design AI file. ArcGIS Pro is the only desktop GIS application that can generate the AIX file, and Maps for Adobe Creative Cloud is the only Illustrator extension that can process these files.

AIX files and Maps for Adobe Creative Cloud

As you will learn later in this chapter, Maps for Adobe Creative Cloud is the tool that performs the file conversion from an AIX file to an AI file. Because the file is converted, you are never actually designing with the AIX file in Illustrator. In addition to sharing AIX files with ArcGIS Pro, ArcGIS users can create AIX files from Map Viewer and ArcGIS® Web AppBuilder in ArcGIS Online. Maps for Adobe Creative Cloud is required to open AIX files shared from these two online applications.

By the end of this chapter, you will be equipped with the necessary background for integrating Maps for Adobe Creative Cloud in your own ArcGIS Pro–to–Illustrator cartography workflows.

About ArcGIS Pro

ArcGIS Pro is a powerful desktop GIS application in which you can create detailed maps using raster and vector spatial data, perform robust analyses of these data, and share work to your ArcGIS Online or ArcGIS Enterprise portal accounts and groups. ArcGIS Pro mapping projects can be in either 2D or 3D and include animation. At the time of this book's publication, the ArcGIS Pro–to–Illustrator workflow is optimized for 2D map creation. For more information about ArcGIS Pro, go to https://pro.arcgis.com.

In addition to ArcGIS Pro being a powerful GIS tool, it is a full map design program. Many cartographers build stunning maps from start to finish by employing the bountiful cartography features offered with ArcGIS Pro. Nevertheless, it is recognized that many mapmakers in the industry complete their final cartographic design in a graphics editor program. It is for this reason that Maps for Adobe Creative Cloud now supports the ArcGIS Pro–to–Illustrator workflow. With this new capability, in addition to creating maps within Illustrator via the extension-direct workflow, ArcGIS Pro users can open their maps and layouts in Illustrator as neatly organized and ready-to-design Illustrator-friendly files.

A complete workflow

Incorporating Maps for Adobe Creative Cloud into your ArcGIS Pro cartography workflow will be explained in two phases: "Part 1, Authoring a map or layout in ArcGIS Pro for Maps for Adobe Creative Cloud" and "Part 2, Working with an AIX file." The first phase covers the steps to properly set up a map in ArcGIS Pro for an optimal AIX export. Opening and working with the AIX file in Illustrator via Maps for Adobe Creative Cloud begins the second phase of the workflow. Users begin with well-organized AI files, complete with structured layers and ready-to-design artwork. Map frames from ArcGIS Pro will load as individual mapboards on the Mapboards panel in Illustrator, allowing mapmakers to add additional layers to maps directly using Maps for Adobe Creative Cloud.

Part 1, Authoring a map or layout in ArcGIS Pro for Maps for Adobe Creative Cloud

Whether you are an ArcGIS Pro expert or new to the application, there are best practices for all mapmakers to consider when authoring an ArcGIS Pro map or layout for an AIX export. This section details important map-authoring considerations to optimize the ArcGIS Pro–to–Illustrator experience.

You can export an AIX file from ArcGIS Pro in one of two ways. The first is to export a map, a quick and simple approach that captures all visible layers in the map extent in the AIX file. The second is to export a layout, which captures the defined page size and any elements added to the page, such as text elements, legends, charts, and multiple map frames.

Required software

For a mapboard to be loaded into Maps for Adobe Creative Cloud correctly, the AIX file must be created using ArcGIS Pro 2.6 or later, and your map or layout must meet the following conditions before exporting the file:

- Turn off Enable wrapping around the date line in the Map Properties window, on the Coordinate Systems tab.

- Confirm that the map frame fits entirely within the layout area.

You will need ArcGIS Maps for Adobe Creative Cloud version 2.2 or later to open AIX files.

Supported map content

The AIX file format supports point, line, and polygon data features, as well as raster layers. Map labels and other map elements—such as legends, titles, and scale bars—are also supported. Because AIX files are intended to be consumed in an Illustrator extension and because Illustrator is a vector graphics editor, vector map data are ideal for the ArcGIS Pro–to–Illustrator workflow.

Point data

In ArcGIS Pro, a point symbol can consist of one or many layers called markers. Each marker can be a shape, picture, 3D, or procedural marker. When your objective includes finalizing a point layer's aesthetic in Illustrator, it is recommended that you use the ArcGIS Pro shape marker. Maps for Adobe Creative Cloud processes shape marker symbols as editable vector paths, making the resulting Illustrator artwork easily editable.

Symbols with multiple layers will be split into separate paths in Illustrator. This includes symbols that comprise shape marker symbol layers (figure 7.1). When using the Processes panel to replace default points with custom symbols, you should use symbols with only one layer. (See "Tutorial 4.2: Custom symbols using the Processes panel" for more information about the Symbols process.)

Fill and stroke colors applied to points in ArcGIS Pro will be preserved in the AIX file, as will transparency. If you still want to style your points as picture markers or 3D markers in ArcGIS Pro, note that picture marker and 3D symbols are processed as image layers in Illustrator.

Figure 7.1. In ArcGIS Pro, a point symbol can consist of many layers called markers. Each marker can be a shape, picture, 3D, or procedural marker. For vector editing in Illustrator, choose shape marker. When the Maps for Adobe Creative Cloud symbols process is part of your workflow, use only single-layered symbols in ArcGIS Pro. In this figure, three layers are used to create one point symbol.

Line data

In ArcGIS Pro, line data, like points, are vectors that can be symbolized with simple symbology or with raster symbology, called strokes. Strokes can be either solid, picture, or gradient. When authoring a map in ArcGIS Pro for an AIX export, use the solid-stroke option for the most flexible Illustrator editing experience. Picture and gradient strokes applied in ArcGIS Pro will be converted into image layers in Illustrator.

Like point symbols, line symbology can comprise multiple layers. If you plan to apply custom brush appearances to your line artwork using the Maps for Adobe Creative Cloud Processes panel, it is best to symbolize line data with a single-layer symbol in Arc-GIS Pro before sharing the AIX file. However, if you choose to keep the line symbology multilayered, make sure that symbol layer drawing is turned on for that layer. See "Mapmaker tip: Symbol layer drawing and categories" in this chapter for more information about this layer setting.

Dash effects on linear data in ArcGIS Pro will be preserved in the AIX file as a dashed-line pattern. Consider figure 7.2, in which the top symbol is a black, 1-point, solid stroke with a dash template of 4 9. This template indicates that the black line will be 4 points, and the space between the black dashes (gap) will be 9 points. When setting a line layer's symbology with a dash pattern, as the top symbol layer indicates in the figure, the dash pattern will be preserved when Maps for Adobe processes the AIX file. Figure 7.3 shows the results of this pattern in Illustrator's Stroke panel. Notice that the stroke pattern is preserved.

If you bring in line data with dash effects that were created in ArcGIS Pro, you will need to select No Constraint At Line Ends (figure 7.4) on the Illustrator

Stroke panel. This option will preserve the line pattern and prevent the resulting line artwork paths from splitting into multiple individual path objects.

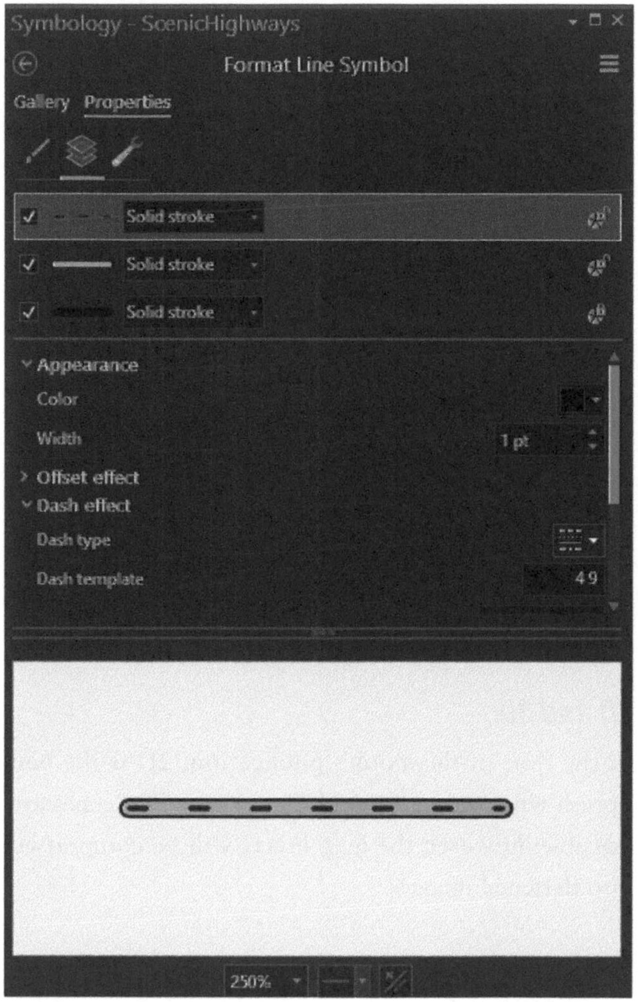

Figure 7.3. The Adobe Illustrator Stroke panel for the dashed line selected. Note the preserved pattern in dash and gap text areas.

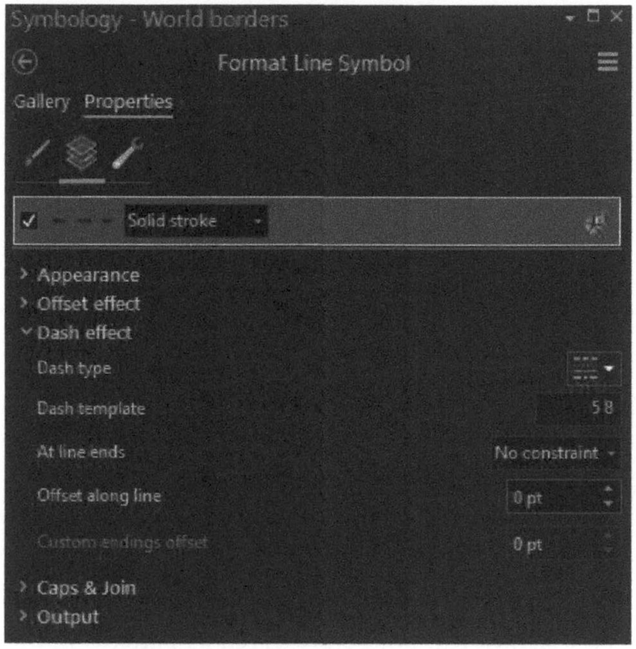

Figure 7.2. Setting up the line style on the ArcGIS Pro Symbology pane, in which the bottom two symbol layers are solid stroke with no dash effect, and the top symbol layer is a solid stroke with a dash effect applied.

Figure 7.4. The Spain population map country borders were set with a line dash pattern of 5-point dashes and 8-point gaps. For the At Line Ends option, No Constraint is selected to preserve this pattern in the AIX file.

The World borders layer of the Spain population map was styled in ArcGIS Pro using a dash pattern of 5 8, or 5-point dashes separated by 8-point gaps. In ArcGIS Pro, this setting can be found in the Dash Effect settings on the Properties tab on the Symbology pane.

Polygon data

Although polygon vector data can be styled in many ways in ArcGIS Pro, the easiest ArcGIS Pro style to work with in Illustrator is the solid-stroke and solid-fill style. Polygon symbology in ArcGIS Pro can have several layers, including picture and 3D marker symbology. Just like points and lines, the recommended practice is to use the simpler symbology.

Pattern-fill styles can be used; however, the pattern will not be preserved in the form of an Illustrator swatch. Instead, the individual shapes that make up the pattern will be processed as individual paths once the AIX file is open in Illustrator, which can create large AI file sizes. Furthermore, these individual paths will be part of a compound path that is clipped by the shape of the geographic polygon. For example, consider the ArcGIS Symbology pane in figure 7.5. The USGS National Hydrography Dataset SwampMarsh category is symbolized by a typical marsh pattern. When an AIX file with this pattern is exported from ArcGIS Pro, the resulting Illustrator layer structure will appear as clipped shapes as shown in figure 7.6. These shapes will not be preserved as patterns in the AIX file. Rather, each individual shape will become its own path.

The outline (*stroke* in Illustrator terminology) of the pattern is its own path (SwampMarsh > <Path> in the layer stack). The patterned artwork of this layer is contained in the <Clip Group> beneath the outline path. Within this clip group is the shape of the feature, which serves as the clipping path for the pattern. Several instances of the pattern are also clipped. If these clipping paths were removed, you would find that each shape that makes up the pattern is itself a path. For this reason, to reduce file sizes and maintain optimal editability in Illustrator, if your polygon layer requires a pattern fill, it is strongly recommended that you add the pattern to the artwork in Illustrator using an Illustrator swatch.

Raster data

All raster data will be brought into Illustrator as image layers. If your ArcGIS Pro map contains multiple raster layers, each individual layer will be preserved. Although raster transparency is supported, it is recommended that you export rasters with their full opacity, and then apply the transparency in Illustrator.

2D and 3D

At the time of this book's publication, 2D is the best option when using the AIX file format. 3D scenes are possible; however, the map layers will be compressed into flattened images.

Basemaps

In ArcGIS Pro, basemaps can be either raster or vector data. Raster basemaps will be processed like all other raster data and brought into Illustrator as image layers. Vector tile basemaps will be processed as vector artwork.

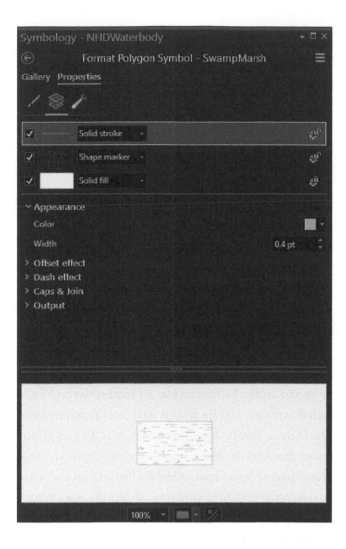

Figure 7.5. In ArcGIS Pro, patterns can be applied to polygon data using the shape marker option. These shapes will not be preserved as patterns in the AIX file. Rather, each individual shape will become its own path. For smaller file sizes and optimal Illustrator editing capability, it is recommended that you use the Swatch panel to apply the pattern to polygon map features in Illustrator.

Figure 7.6. The ArcGIS Pro symbology from figure 7.5, *left*, when opened in Illustrator, *right*, results in a series of clipped shapes instead of an Illustrator swatchlike pattern. For better editability, style polygons with solid fill in ArcGIS Pro, and apply patterns in Illustrator using the Swatch panel.

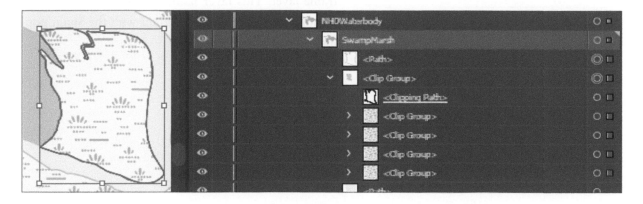

Simplify complex data for ease of rendering in Illustrator

GIS data with a lot of vertices lead to big file sizes. Often, having many vertices is useful for showing appropriate detail at the large scale, but they become superfluous at a smaller scale. To render complex data with a lot of features and vertices, ArcGIS Pro splits the data rendering into pieces. In ArcGIS Pro, this does not impact the appearance. However, when a complex dataset is split up and then opened in Illustrator, editing the layer's style can be more complicated than necessary. For example, a country with an extremely complex shoreline might have its shoreline split into multiple paths when you open the map in Illustrator. You would then need to merge these paths using Illustrator's Pathfinder tools. In these circumstances, before exporting from ArcGIS Pro, consider using a data management tool to reduce the number of vertices. Before exporting these complex data from ArcGIS Pro, consider using a data simplifying tool from the Generalization toolset, at links.esri.com/Generalization.

Map layers

This section describes optimal ways to organize ArcGIS Pro map layers for an ideal Illustrator layer structure. Following this discussion, you will be presented with the Spain population map as it was set up in ArcGIS Pro and a comparison of the same map when opened as an AIX file in Illustrator using Maps for Adobe Creative Cloud.

Layer names and group layers

When an AIX file is processed using Maps for Adobe Creative Cloud, the resulting AI file's layer structure will mirror the ArcGIS Pro layer structure. This structuring includes layer names and group names. If you are sharing from an ArcGIS Pro layout that has a single or multiple map frames, these map frames—and their names—will also become part of the AI file's layer structure. In Illustrator, all feature layers' associated artwork will be placed into their respective layers. If these layers were part of an ArcGIS Pro group layer, the ArcGIS Pro group layer becomes an Illustrator parent layer containing the feature layers. This parent layer will be a sublayer within the map frame layer. Also note that if a feature layer was not placed in a group layer, it will be a direct sublayer of the map frame layer. This layer naming and structure will be demonstrated later in this chapter with the Spain population map.

Embedding fonts

In addition to embedding fonts from your map labels and elements, the Embed Fonts option preserves special characters and diacritic marks used in the map. When exporting an AIX file, make sure that font embedding is enabled in ArcGIS Pro if any layer names contain diacritic or non-Latin characters (figure 7.7). This will help preserve the appropriate characters in the resulting AI file.

˅ Fonts
☑ Embed fonts
☐ Convert character marker symbols to polygon

Figure 7.7. When ArcGIS Pro layer names have any diacritic or non-Latin characters, checking Embed Fonts ensures that the characters are preserved in the Illustrator layer names.

Classed and categorized layers

For any layer in an ArcGIS Pro map or layout that is classified either quantitatively or qualitatively, the resulting layer in Illustrator will contain sublayers for those classes. These classed sublayers will be nested within the feature layer. For example, a point dataset named Roads that has been classified in ArcGIS Pro by its type category—in which the dataset contains a local, arterial, and highway type—will result in an AI file with a Roads layer containing sublayers named local, arterial, and highway. See "Mapmaker tip: Symbol layer drawing and categories" for optimal layer structure workflows.

Definition queries filters

All features that are visible and within the ArcGIS Pro layout or map extent will be present in an AIX file's resulting AI file. Features that have been hidden—or filtered out—by a definition query in ArcGIS Pro will not appear in the AI file.

Projections

A map's projection in ArcGIS Pro will be processed in the AIX file, and the resulting Illustrator file will also be in the same projection. Note that at the time of this book's publication, if an AIX file is created using a non-Web Mercator projection, the Maps for Adobe Mapboards panel will display only Web Mercator, although the Compilation panel will reflect the correct projection.

Symbol layer drawing and categories

To make sure your map layer's categories (or classes) appear in the identical order in Illustrator as they do in ArcGIS Pro, enable the layer's symbol layer Drawing setting. To do this, in ArcGIS Pro, open the Symbology pane for the desired layer, and click the Symbol layer Drawing tab. Then toggle on Enable Symbol Layer Drawing. With this setting turned on, you can manually reorder the categories to logically fit the layer's needs. For example, placing highways on top of local streets will ensure that the highway features' artwork is above local streets in Illustrator. If the Enable Symbol Layer Drawing option is not toggled on, the category sublayers may not match the expected stacking order.

The following two figures show the drawing order for a choropleth layer that shows the population density for Madrid's municipality regions. In figure 7.8, the ArcGIS Pro Symbology pane displays the Symbol Layer Drawing tab as selected. Notice that Enable Symbol Layer Drawing is toggled on. Figure 7.9 demonstrates the Illustrator file's layer order for these categories, which matches the order indicated in ArcGIS Pro.

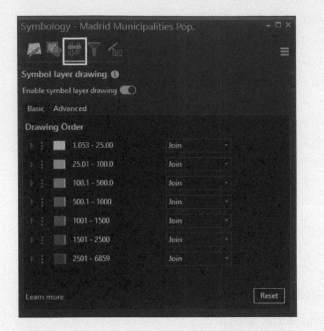

Figure 7.8. To ensure category sublayers in Illustrator will be in the expected order, Enable Symbol Layer Drawing is toggled on for the layer in ArcGIS Pro.

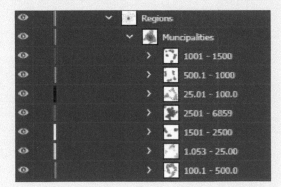

Figure 7.9. When Enable Symbol Layer Drawing is toggled on for Madrid Municipalities Pop. in ArcGIS Pro, the resulting Illustrator layer order will match the order of categories from ArcGIS Pro.

Symbology for the ArcGIS Pro-to-Illustrator cartographer

The AIX file type supports most symbology settings from ArcGIS Pro. However, as you have discovered thus far in this chapter, best practices exist. Here are some additional tips to consider when authoring an ArcGIS Pro map or layout for AIX file export:

- If you want to perform additional graphic editing on your Illustrator vector artwork, at the time of this book's publication, the most flexible ArcGIS Pro symbology types are Single Symbol, Unique Values, Graduated Colors, and Graduated Symbols. Other symbology types, such as Bivariate Colors and Unclassed Colors, are also supported.

- Transparency settings from ArcGIS Pro are transferred to the AIX file. Any transparency applied to a group layer will also be applied to that group's nested layers. If a group's individual layers require different transparency settings, it is best to set the transparency for the layer itself rather than the group. Transparency can also be applied to artwork once the map is open in Illustrator.

Text

Labels

Each map or map frame in ArcGIS Pro that contains a labeled layer will have a parent layer named Labels in the resulting AI file. A sublayer within the Labels layer will be created for any labeled layer. The name of these sublayers will match the layer name to which the labels belong. For example, figure 7.10 shows the Illustrator layer structure for the Spain population map's main map frame, Main Map – Spain, and the labels for that map frame. The Labels parent layer contains one sublayer for each of the labeled layers in the Main Map – Spain map frame. This structure ensures that all labels are above the map features and are therefore visible on the map.

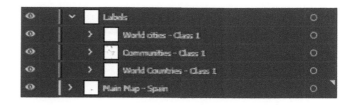

Figure 7.10. Because the map frame for Main Map – Spain consisted of three layers that were labeled in ArcGIS Pro, a parent Labels layer is added to the AI file above the parent Main Map – Spain layer. Within this Labels layer are three sublayers, one for each layer that was labeled. The appended " – Class 1" to each sublayer is the name of the ArcGIS Pro label class.

Halos and outlines

ArcGIS Pro text halos and text with outline symbology will result in non-text paths in Illustrator. Consider the label for Beauty Mountain. In ArcGIS Pro (figure 7.11, top), this label uses the Tahoma font and was styled as a white label with a black halo. In Illustrator, although the label is preserved as editable text in the original font, the halo is instead a shape and not editable as text. If you want to make changes to these labels, such as editing the font, kerning, point size, and other changes, the halo will no longer be suitable for the label (figure 7.11, bottom). If you are planning to edit text once you open your map in Illustrator using Maps for Adobe Creative Cloud, it is recommended that you apply text halos and outline symbology (strokes) in Illustrator (figure 7.12).

Beauty Mountain

Beauty Mountain

Figure 7.11. The Beauty Mountain label, *top*, shows its appearance when styled with the Tahoma font, in white with a black halo, in ArcGIS Pro, and then opened in Illustrator. The white part of the label is editable text in Illustrator, yet the black halo is no longer text. Thus, updating the label's appearance in Illustrator will not be reflected in the halo. The bottom label shows the results when the label's font and point size are edited in Illustrator. Because font and point size are text characteristics, they cannot be applied to the black halo.

BEAUTY MOUNTAIN

Figure 7.12. In this example of the Beauty Mountain label, the halo was applied in Illustrator after the label's font, point size, and tracking were edited.

Fonts

For fonts from an ArcGIS Pro AIX file to appear in Illustrator, in addition to turning on font embedding on the Export Map pane, you must install the font on the machine used to open the AIX file. Any missing fonts will be replaced with a default font.

Things to know when authoring an ArcGIS Pro layout

In ArcGIS Pro, you have the option to export AIX files from a map or a layout. An ArcGIS Pro page layout—often simply referred to as layout—can consist of one or several maps and elements, such as multiple map frames, charts, scale bars, descriptive text, and titles. Layouts are fantastic for arranging map elements in an organized fashion for print. Exporting an AIX file from a layout view can be incredibly useful for mapmakers who continue their workflow in Illustrator. Considerations and recommendations for exporting your ArcGIS Pro layouts as AIX files are described in this section.

Map frames

You can export ArcGIS Pro layouts that have single or multiple map frames. After you open the AIX file using Maps for Adobe Creative Cloud in Illustrator, the resulting AI file will have a parent layer for each map frame. The two map frames for the Spain population map, Inset map – Madrid and Main Map – Spain (figure 7.13), will be parent Illustrator layers when exported as an AIX file. The individual layer names will be visible in those layers. Moreover, each map frame will load into Maps for Adobe Creative Cloud as an individual mapboard on the Mapboards panel, which allows users to add new map data directly from the extension. For the Spain population map, there will be a mapboard for each of the Inset map – Madrid and Main Map – Spain map frames. You will learn more about mapboards and AIX files in the "Part 2, Working with an AIX file" section later in the chapter.

Layout elements

Layout elements such as descriptive text, graphic elements, and legends that are included in an AIX file export will be processed as editable vector objects. Note that any image elements in a layout will also be processed as images in an AIX file export.

Figure 7.13. The ArcGIS Pro page layout for the Spain population map. The map frames have been collapsed on the Contents panel so their respective layer names are not visible in the figure. Notice the two map frame names, Inset map - Madrid and Main Map - Spain. These two map frames will serve as parent layers in Illustrator, containing all the feature layers for each frame.

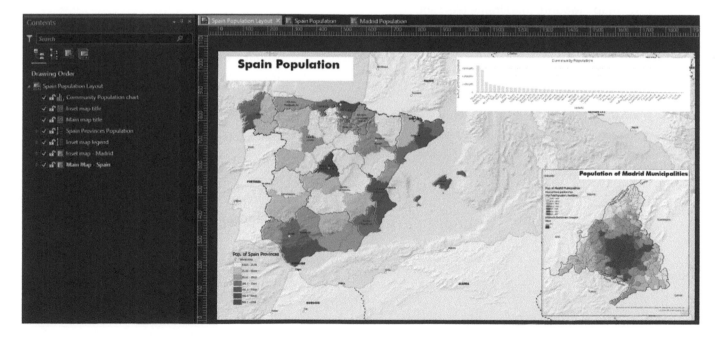

Spain population map: ArcGIS Pro layout layer structure

Figure 7.14 shows the Spain population map's Contents pane expanded, exposing the layer drawing order in ArcGIS Pro. Note that some of these layers' visibility is turned off, which means they will not be part of an exported AIX file. The order is explained as follows, from top to bottom.

- **Spain Population Layout:** The name of the map layout.

- **Community Population chart:** A chart frame containing a bar chart showing the population of Spain's communities.

- **Inset map title and main map title:** Text elements containing titles for the respective map frames.

- **Spain Provinces and Madrid Municipality Legends:** A legend for each map.

- **Inset map - Madrid:** Map frame element containing the Madrid Population map frame (see the tab for this map in figure 7.13), including its data layers described in the following map layers point.

- **Main Map - Spain:** Map frame element containing the Spain population map, including its data layers described in the map layers point.

- **Map layers:** Both map frames contain identical map layers. However, the layer visibility differs slightly between the two map frames. Remember, when a layer's visibility is checked off, the layer will not become part of the AIX file.

- **World cities:** Point data layer symbolized by a shape marker symbol. This layer's visibility is toggled off in the inset map.

- **World borders:** Line data layer depicting the boundaries of world countries.

- **Spain Regions:** A group layer containing the following data layers:

  - **Communities:** Polygon data layer containing the autonomous communities of Spain. This layer is symbolized by a solid gray stroke and no fill color.

  - **Madrid Municipalities Pop.:** Polygon data layer containing the municipalities within the Community of Madrid. This layer is symbolized as a choropleth depicting the population of communities. Each category's symbology has a 25 percent transparency. To ensure proper drawing order in the AIX file, symbol layer drawing is enabled for this layer (see "Mapmaker tip: Symbol layer drawing and categories").

- **Provinces:** Polygon data containing Spain's provinces. In the Main Map – Spain map frame, this layer is symbolized as a choropleth depicting the population of provinces. Each category's symbology has a 25 percent transparency. Symbol layer drawing is enabled. The Provinces layer in the Inset map – Madrid map frame has a display filter removing the Madrid province.

- **World Countries:** Polygon data layer of world countries, symbolized with a simple gray file and simple stroke outline. This layer in the Main Map – Spain map frame has a data filter that removes the country of Spain.

- **Oceans:** Polygon layer of world oceans.

- **Hillshade-Bathymetry Imagery:** Raster layer showing the physical topography of Earth's surface.

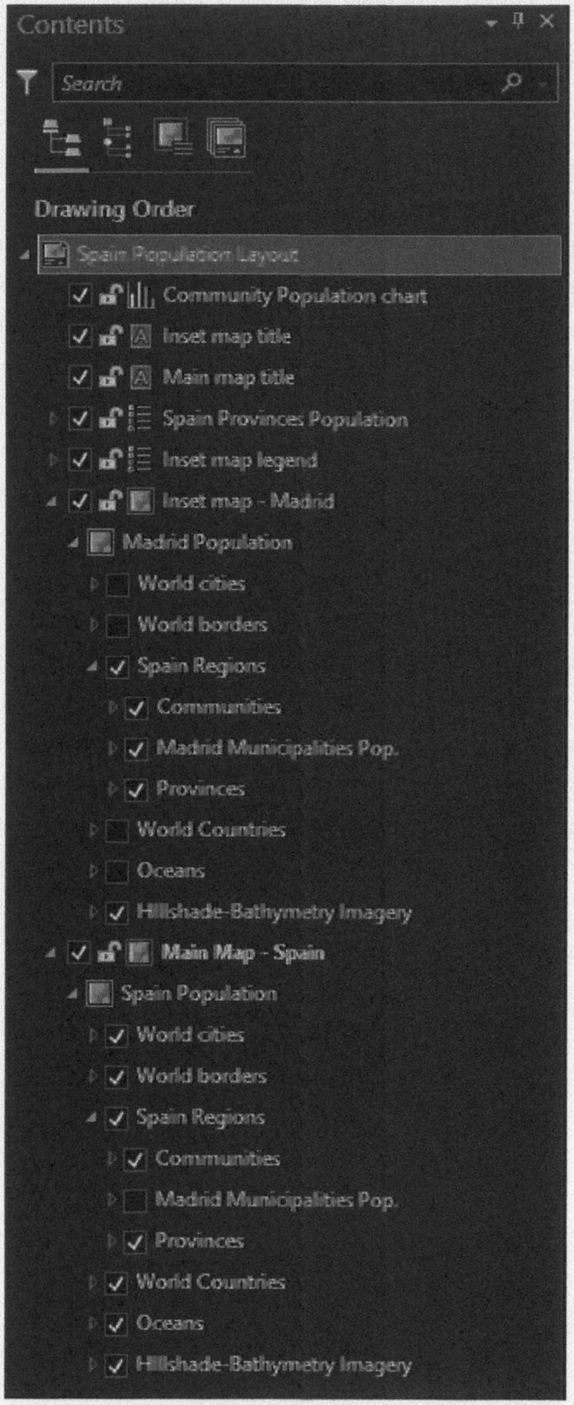

Figure 7.14. The ArcGIS Pro Contents pane for the Spain population map's page layout. Some layers are unchecked, indicating that they will not be included in the AIX file.

Exporting from ArcGIS Pro to Illustrator via AIX file

The main difference between exporting a map and a layout is the way in which the geospatial data are viewed. A map in ArcGIS Pro is simply the display of map layers that have been added to the table of contents. Maps can be either 2D or 3D, or they can be basemaps. 3D maps are called scenes. As previously specified, layouts can have several map elements, including multiple map frames. This is in contrast with maps, which contain only the layers visible in the table of contents. For more information on ArcGIS Pro page layouts, see links.esri.com/Layouts.

Sharing maps and layouts

After a map is created in ArcGIS Pro, where map layers have been added, analyzed, and symbolized, you can follow one of the two approaches for exporting an AIX file. Whether exporting an AIX file from a map or a layout, each approach has its own set of best practices. Both approaches are accomplished by choosing the appropriate settings on their respective Export panes, accessed from the Share tab in ArcGIS Pro.

Sharing a map

To share an ArcGIS Pro map as an AIX file, you must first have an open and active map. Make sure that all the layers you want in your AIX file are visible on the Contents pane and that all layers have been symbolized and labeled (if desired).

1. On the ribbon, click the Share tab, and in the Output group, click the Export Map button to open the Export Map pane.

2. With the Export Map pane open, enter your desired AIX file settings, described in the "Export Map pane settings" sidebar. Once finished, click Export to create the AIX file.

3. Open the AIX file either by clicking the View Exported File link at the bottom of the Export Map pane or by signing in to Maps for Adobe Creative Cloud and opening the AIX file in Illustrator.

For the View Exported File method to function properly, you must first be signed into Maps for Adobe Creative Cloud. For this part of the AIX file workflow, see the "Part 2, Working with an AIX file" section later in the chapter.

Export Map pane settings

The ArcGIS Pro Export Map pane settings (figure 7.15) are as follows:

- **File Type:** The first option in the File section of the pane is the File Type drop-down list. Select AIX to export a map of this format.

- **Name:** The second option is the Name field. This field includes both the name and location to which the AIX file will be saved. The map name is automatically applied as the default name; however, you may customize the name to fit your needs. Note that this name will be applied only to the AIX file. Once the file is open in Illustrator, it will be converted to an AI file and will need to have a name chosen. The name of the AIX file is not transferred.

- **Compression:** The Compression section's settings include a setting for image compression and one for vector graphics compression. The image compression drop-down list contains options for the compression scheme used to compress image or raster data in the output file. The image compression options are None, Run-Length Encoded (RLE), Deflate, Lossless (LZW), JPEG, and Adaptive. The Quality option is activated when any image compression option other than None is chosen. A lower quality will result in a smaller file size with less crisp image output, and higher quality will result in a larger file size with crisper image output. When the Compress Vector Graphics check box is checked, the content vector streams will be compressed.

- **Image Size:** The map view determines the extent and scale of the exported map. However, the size can be updated in the image size settings. Although this setting has *image* in the name, it is the setting for the entire final output size of the AIX file. When editing the size setting, click the Preserve Aspect Ratio button ⚙ when you need to lock this parameter. Then update the size manually, either by typing directly in the Width and Height text fields or by clicking the arrows associated with the width and height.

- **Resolution:** This setting determines the amount of image resampling of raster content in the map export. For example, if the resolution is set to 300 dots per inch (DPI), and the ratio is set to 1:2, the resulting raster content's DPI will be 150. Higher raster resolution will result in larger file sizes.

- **Fonts:** With the Embed Fonts box checked, any embeddable fonts will be included in the AIX file. Note that for the font to appear in the resulting AI file, the machine used to open the AIX file must have those fonts installed.

ArcGIS Pro provides many Esri fonts that can be used as map symbols. If you would like to keep the appearance of these symbol characters and are unsure if the machine you are using to open the AIX file has these Esri fonts, check the box for Convert Character Marker Symbols To Polygon. This will convert the symbol characters to outlines instead of preserving them as text.

Color Management: With the Embed Color Profile box checked, the map's color profile will be included in the AIX file. This setting will help preserve color consistency across devices.

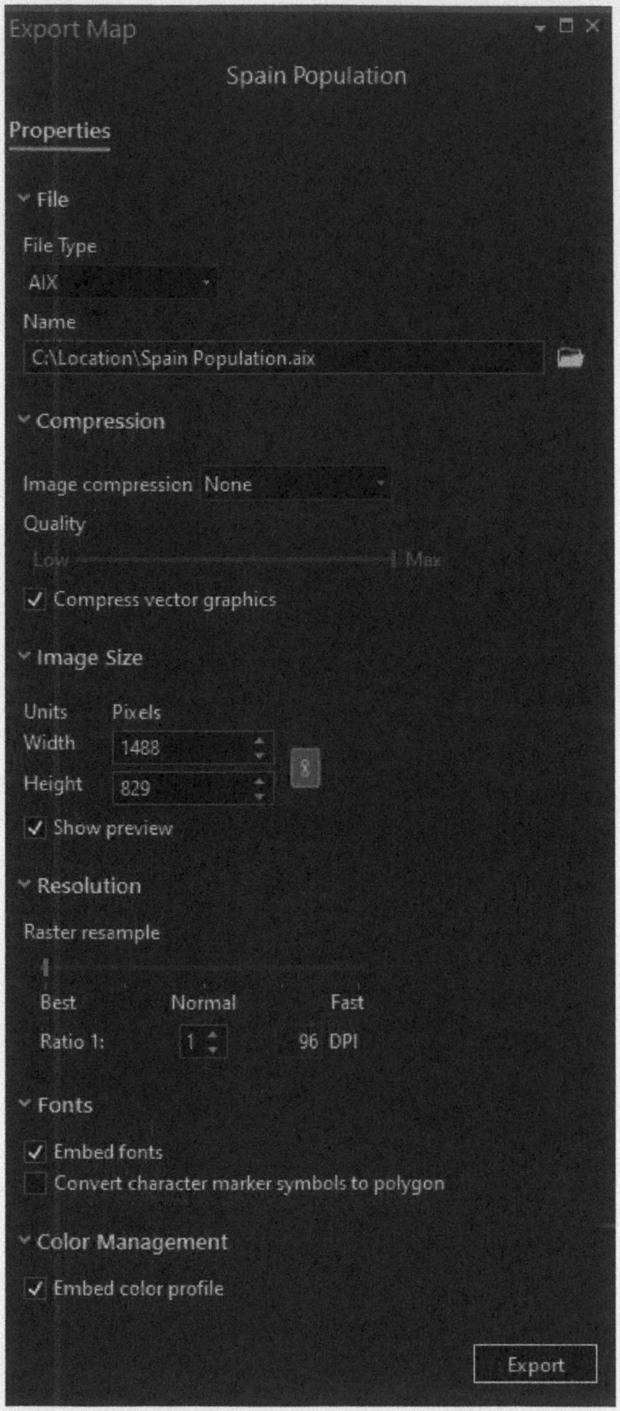

Figure 7.15. The ArcGIS Pro Export Map pane.

Mapping by Design: A Guide to ArcGIS Maps for Adobe Creative Cloud

Show preview when editing output size

When exporting maps from ArcGIS Pro, if you edit the image size setting and choose to not preserve the aspect ratio, the map may become cropped. For example, if your map view is 1000 × 750 pixels and you changed the export size to 500 × 500, the content is cropped on the left and right sides. This is because the registration of the image size is centered on the map. Check the box next to Show Preview to view any cropping that may occur. Note that the preview occurs on the map itself and not on the Export Map pane.

Sharing a layout

When sharing an ArcGIS Pro layout as an AIX file, you must have an open and active layout that contains at least one map frame. Just as when exporting an AIX file from a map, when sharing a layout make sure that all the layers you want in your AIX file are visible on the ArcGIS Pro Contents pane. Also make sure to add any desired labels and filters and perform any data analyses before exporting.

1. To export the layout, click the ribbon's Share tab, and then click the Export Layout button to open the Export Layout pane.

2. With the Export Layout pane open, enter your desired settings described in the "Export Layout pane settings" sidebar. Then click Export to create the AIX file.

3. Open the AIX file either by clicking the View Exported File link at the bottom of the Export Layout pane or by signing in to Maps for Adobe Creative Cloud and opening the AIX file in Illustrator.

For the View Exported File method to function properly, you must first be signed in to Maps for Adobe Creative Cloud.

Export Layout pane settings

On the Export Layout pane (figure 7.16), the Compression, Fonts, and Color Management settings follow the same behavior as on the Export Map pane. Rather than reiterate the settings that are identical between the two, this section covers only the settings on the Export Layout pane that either do not exist in, or differ from, the Export Map pane.

- **Clip To Graphics Extent:** On the Export Layout pane, the behavior of the file type and name settings in the File section's settings matches that of the Export Map pane. There are two additional features on the Export Layout pane: Clip To Graphics Extent and Keep Layout Background. When the first one is checked, the output will be cropped to the extent of the map elements rather than the extent of the page.

  Figure 7.17 shows the ArcGIS Pro page layout before export, and figure 7.18 shows the Illustrator output that will result when Clip To Graphics Extent is checked for this page layout. Notice that the two map frames have defined the AIX file's extent rather than the page layout. When Clip To Graphics Extent is left unchecked, the output in Illustrator will match that of the page layout settings.

- **Keep Layout Background:** When this box is checked, the output in Illustrator will have a background layer at the bottom of all other layers in the file. This background layer will have a white rectangular path that matches the size of the artboard.

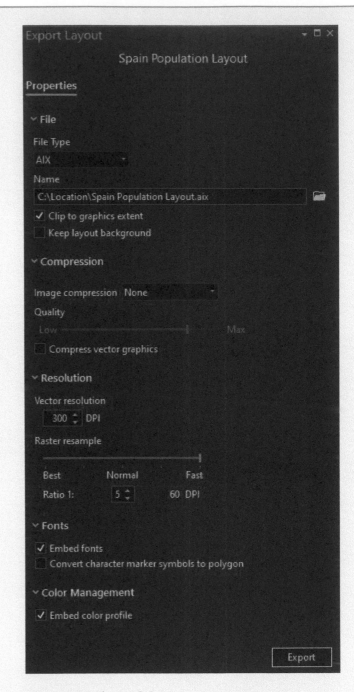

Figure 7.16. The ArcGIS Pro Export Layout pane.

- **Vector Resolution:** This setting, available only on the Export Layout pane, determines the DPI of the vector data in the AIX file. Vector resolution is automatically calculated when exporting a map.

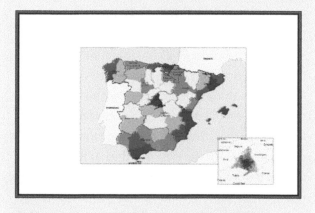

Figure 7.17. The ArcGIS Pro layout with two map frames.

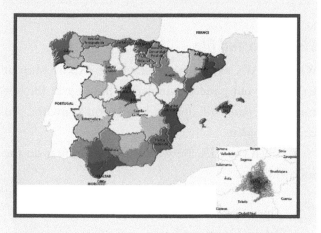

Figure 7.18. When Clip To Graphics Extent is checked in the ArcGIS Pro Export Layout pane, the results in Illustrator will be cropped to the extent of the elements within the ArcGIS Pro layout.

ArcGIS Pro-to-Illustrator one-way workflow

At the time of this book's publication, exporting AIX files from ArcGIS Pro is a one-way workflow. Returning your map layers to ArcGIS Pro is not yet available. You can, however, export a web map from Maps for Adobe Creative Cloud to ArcGIS Online, and then open it in ArcGIS Pro.

Sometimes, mapmakers will discover that new data or analyses are required after they open an AIX file in Illustrator. Or you might find that the layout and map extent need to be changed. In these cases, update the map project as needed in ArcGIS Pro and share a new AIX file.

Part 2, Working with an AIX file

After you export your map or layout as an AIX file, you can open it in Illustrator once you have signed in to the Maps for Adobe Creative Cloud extension. When you open the AIX file, Maps for Adobe Creative Cloud will convert the file into a new AI file that consists of well-organized layers containing the map's features, labels, and elements. Any map frame from the AIX file will be loaded into the Mapboards panel as a mapboard, to which new data layers from ArcGIS Online or your computer can be added. These mapboards are created using the properties of your AIX file, including the spatial extent, scale, and map projection. This section details the workflow and expectations for opening an AIX file, as well as information for adding new map features to the resulting AI file.

Opening an AIX file

AIX is an interchange file type. As stated earlier in this chapter, you can think of opening an AIX file as a conversion process in which Maps for Adobe Creative Cloud is the tool that performs the file conversion from an AIX file to an AI file. For this reason, you must be signed in to Maps for Adobe Creative Cloud to open an AIX file. Instead of designing with the AIX file in Illustrator, immediately after Maps for Adobe Creative Cloud processes the AIX file, you will have an untitled AI file.

Save the AI file as soon as your AIX file is open

When Maps for Adobe Creative Cloud processes an AIX file made with ArcGIS Pro, the name of the AIX file does not persist in the resulting AI file. Instead, once an AIX file is open, the new AI file is titled Untitled-1.ai, in which the number is serialized chronologically by the number of AIX files opened during one session. For example, if you open three AIX files during one Illustrator session, the AI file names upon opening will be Untitled-1.ai, Untitled-2.ai, and Untitled-3.ai, respectively. If the new untitled AI file is closed before saving, it cannot be recovered. It is strongly recommended that you save the untitled AI file with a new name immediately upon opening it.

View exported file

When an AIX export from either the Export Map or Export Layout pane is complete, a message at the bottom of the pane will appear, letting you know the process is complete.

To open the AIX file with this option, first do the following:

1. Open Illustrator, and sign in to Maps for Adobe Creative Cloud.

2. From the respective Export pane, click View Exported File.

An ArcGIS Maps for Adobe Creative Cloud splash screen appears (figure 7.19), letting you know that Maps for Adobe Creative Cloud is opening and processing your AIX file. The process duration is determined by the AIX file size.

Figure 7.19. When opening an AIX file, the ArcGIS Maps for Adobe Creative Cloud splash screen appears to let you know that the file is being processed.

Opening an AIX file from Illustrator

The second way to open an AIX file is directly from Illustrator.

1. Open Illustrator, and sign in to Maps for Adobe Creative Cloud.
2. From the Illustrator File menu, click Open.
3. Locate the AIX file, and select it.

You will see the Maps for Adobe splash screen as the extension processes the AIX file.

Expected Illustrator layer structure

Illustrator layer structure from map exports

This section describes the Illustrator layer structure when a map (as opposed to a layout) is shared as an AIX file export, using the Spain population map as an example. Figure 7.20 shows the map in ArcGIS Pro, and figure 7.21 shows this map's layers in Illustrator. Refer to the "Spain population map: ArcGIS Pro layout layer structure" sidebar earlier in this chapter for how these layers were designed, including transparency, symbol layer drawing, and data filters.

- **Map parent layer (map exports):** A map AIX file export will have a parent layer containing all the geographic feature layers as sublayers. The name of this map parent layer matches the name of the map in ArcGIS Pro. In figure 7.20, the tab shows that the ArcGIS Pro map is named Spain Population as is the Illustrator parent layer for this map.

- **Labels parent layer (map exports):** As described in the previous section, if your map has labels, a parent layer named Labels will be created that contains a sublayer for each labeled layer. These sublayers' names will match the feature layer to which they belong. In the case of the Spain population map, the " – Class 1" is appended to these layer names representing the label class name that was used in ArcGIS Pro. Note that with a custom label class name, the layer name will be appended with the custom name—for example, World cities – custom label class name.

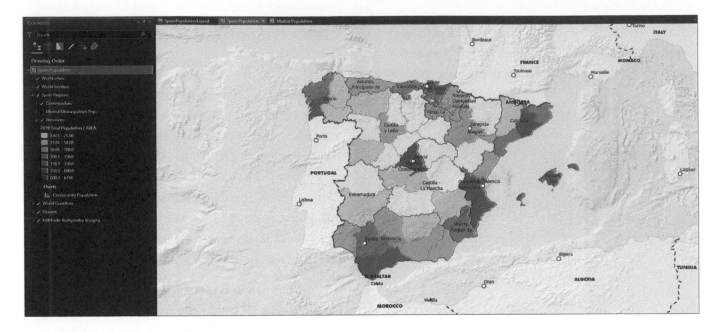

Figure 7.20. The map of Spain in ArcGIS Pro. This is not the page layout but only the map itself.

- **Sublayers for geographic features (map exports):** When you export a map, the main map parent layer will have a respective sublayer for each data layer from the original ArcGIS Pro map. These first-order sublayers will be named identically as their respective data layers from ArcGIS Pro.

- **Illustrator layer results from ArcGIS Pro group layers (Map exports):** If the data layers were grouped in ArcGIS Pro, the main map parent layer will have a first-order sublayer representing the group, and this first-order sublayer will have its own set of sublayers—one sublayer for each data layer that was part of the ArcGIS Pro group. The group-level layer organization and layer visibility are demonstrated in figure 7.21,

in which the Illustrator main map parent layer, Spain Population, has a first-order sublayer called Spain Regions with two sublayers: Communities and Provinces.

- **Illustrator layer results from layer categories and classes (map exports):** When a data layer is classed or categorized in ArcGIS Pro, the Illustrator sublayer that is created for the data layer will have its own set of sublayers that represent the categories. Spain's provinces were symbolized as a choropleth layer classified into groups that represent their population density. The Illustrator Layers panel shows the following hierarchy for these data classes: Spain Population > Spain Regions > Provinces > seven sublayers, with one for each choropleth class.

Mapping by Design: A Guide to ArcGIS Maps for Adobe Creative Cloud

Illustrator layer structure from layout exports

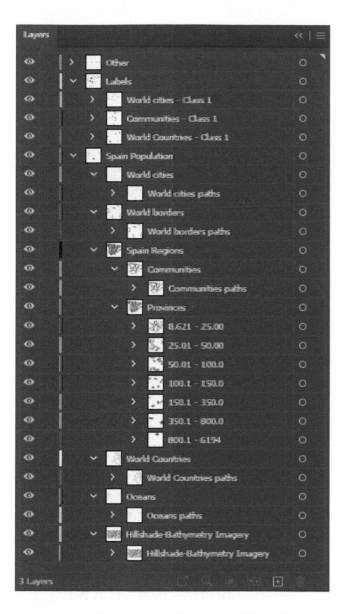

Figure 7.21. The Illustrator layer panel from the AI file that results from an ArcGIS Pro AIX file map.

The resulting Illustrator layer structure from an Arc-GIS Pro layout export differs from the map export layer structure in a few ways. This section describes the similarities and differences in the ArcGIS Pro layout to Illustrator using the Spain population map (see figures 7.13 and 7.14 for the original layout in ArcGIS Pro) as an example. The ArcGIS Pro contents for this layout can be seen in the "Spain population map: Arc-GIS Pro layout layer structure" sidebar. The resulting AI file's layer structure is shown in figure 7.22.

- **Element layers (layout exports):** Each map element will be organized into its own discrete Illustrator layer and will appear in the same order as on the ArcGIS Pro Contents pane. In the "Spain population map: ArcGIS Pro layout layer structure" sidebar, you will notice that each map element in the Spain population map is at the top of the layer stack on the ArcGIS Pro Contents pane, and they are not within a group layer. The resulting Illustrator layer structure matches this stacking order, in which each element's artwork is placed into a discrete layer with a name that matches the ArcGIS Pro element. In figure 7.22, there are five layers representing each of the map elements in the Spain population map: a chart layer, two text element layers—one for each map title—and two layers representing legends for the main and inset maps.

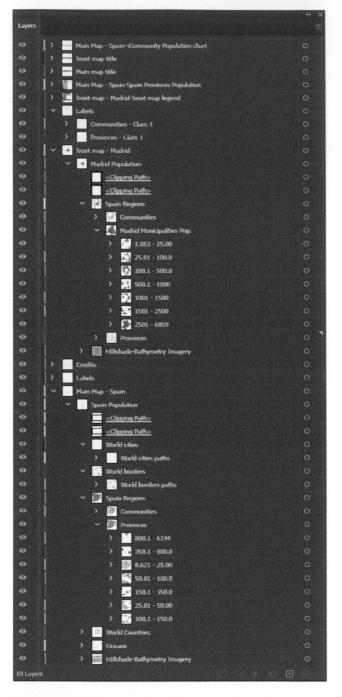

Figure 7.22. The resulting Illustrator layers for an ArcGIS Pro page layout shared as an AIX file.

- **Labels layers (layout exports):** There will be one parent layer named Labels for each map frame that contained a layer that was labeled. These Labels layers will have a sublayer for each layer that was labeled in the map frame. These sublayers' names will match the feature layer to which they belong. As you can see in figure 7.22, there is a Labels parent layer above the Inset map – Madrid parent layer. Nested in this Labels layer are two sublayers, one for each layer that was labeled in ArcGIS Pro: Spain's Communities and Provinces. Just as in map exports, class names are appended to these Illustrator layer names representing the label class that was used in ArcGIS Pro.

  Notice that there is also a Labels parent layer above the Main Map – Spain parent layer. Sublayers for each labeled feature layer are nested in this second Labels layer; the parent layer is collapsed, hiding these sublayers from view.

- **Sublayers for geographic features (layout exports):** When a layout is exported to an AIX file, the map frame parent layers in Illustrator will have a respective sublayer for each data layer from the original ArcGIS Pro map. These sublayers will be named identically as their respective data layers from ArcGIS Pro.

- **Illustrator layer results from ArcGIS Pro group layers (layout exports):** The expectation of group layers in an ArcGIS Pro layout AIX file export is the same as a map AIX file export. Because layouts can have map elements, these elements may also be placed in group layers.

- **Illustrator layer results from layer categories and classes (layout exports):** In ArcGIS Pro layout exports, data layer categories and classes follow the same Illustrator layer structure as in map exports.

Mapboards panel: Syncing AIX file exports to Maps for Adobe Creative Cloud

The Mapboards panel and map exports

Just as when maps are created using Maps for Adobe Creative Cloud and the extension-direct workflow, AIX files are synced to the extension. There are a few differences in the behavior of a synced map created from an AIX file. When you open an AIX file that was exported from an ArcGIS Pro map, a mapboard that matches the original map's extent will load on the Maps for Adobe Creative Cloud Mapboards panel (figure 7.23). Note that the extent of the mapboard will match whatever the ArcGIS Pro map extent was at the time of export. Although the AIX file name does not persist in the resulting AI file, the name is kept in the mapboard name.

The Mapboards panel and layout exports

When an ArcGIS Pro page layout is shared as an AIX file and then opened in Illustrator, a mapboard for each ArcGIS Pro map frame will be loaded in the Maps for Adobe Creative Cloud Mapboards panel. In the case of the Spain population map, two map frames are loaded: a Spain Population_Main Map – Spain mapboard and a Madrid Population_Inset

map – Madrid mapboard. The names for these mapboards are derived from combining the ArcGIS Pro map name and map frame name (Map Name_Map Frame Name). Users can select which mapboard to focus on in the Mapboards panel's mapboard selector (figure 7.24).

Compilation panel expectations: Linking AIX file exports to Maps for Adobe Creative Cloud

You can use Maps for Adobe Creative Cloud to add new data to your AIX file–generated AI files. This is performed in the same manner that you learned in previous chapters by adding data to the mapboard. These new layers can be local files from your computer, as well as layers or web maps hosted in ArcGIS Online.

MAPMAKER TIP

Use the Compilation panel for adding new layers

Make sure you add new data to an AIX file–generated mapboard from the Compilation panel. If data are instead added from the Mapboards panel, the original map extent will be overridden by the extent of the added data.

Just as you have learned throughout this book, in Maps for Adobe Creative Cloud, new data are typically added from the Compilation panel (see the mapmaker tip). Unlike the extension-direct workflow, in which all layers from the most recent sync appear on the Contents panel, in the ArcGIS Pro–to–Illustrator

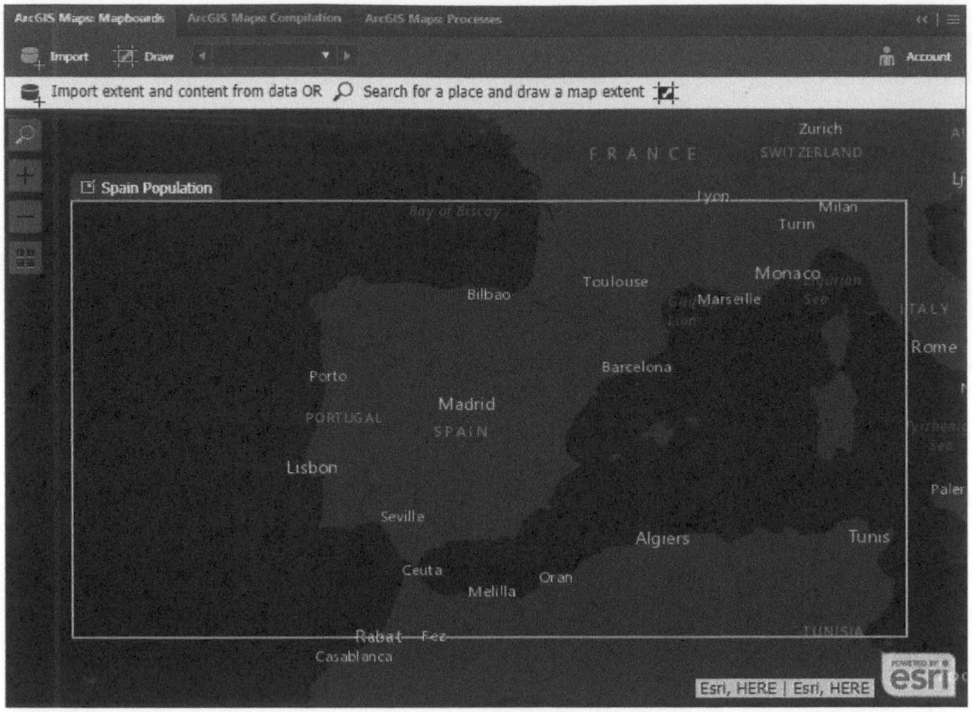

Figure 7.23. After the Spain population map is shared as an AIX file, the map extent will load as a mapboard on the Maps for Adobe Creative Cloud Mapboards panel. This allows you to continue adding more map layers using the extension.

Figure 7.24. After the Spain population map layout is shared as an AIX file, a map extent for each map frame is loaded as a mapboard on the Maps for Adobe Creative Cloud Mapboards panel.

Mapping by Design: A Guide to ArcGIS Maps for Adobe Creative Cloud

workflow, the Maps for Adobe Creative Cloud Contents panel will not show layers that were added in ArcGIS Pro. Likewise, these layers will not appear in the map preview area. Whereas the Contents panel and map preview area appear blank, the mapboard name will appear at the bottom of the Compilation panel. Figure 7.25 shows that the Spain Population_ Main Map – Spain mapboard is active.

Figure 7.26. To provide spatial context, a World Countries layer is added to the Compilation panel.

Figure 7.25. The Compilation panel upon opening the Spain population map AIX file export. This view shows Spain Population_Main Map - Spain as the selected mapboard.

The next section describes the steps for adding new data to an AIX file–generated mapboard using the Spain population map as an example. To give the Compilation panel visual spatial reference for readers, a World Countries layer has already been added to the mapboard with Maps for Adobe Creative Cloud using the ArcGIS Online library (figure 7.26). With the addition of this layer, it becomes clear that the mapboard extent matches the ArcGIS Pro layout export.

Adding new data to an AIX file-generated map via Maps for Adobe Creative Cloud

1. On the Compilation panel, click Add Content, and select local files, places, or ArcGIS Online layers.

You can still add data to your AIX file–generated maps using Maps for Adobe Creative Cloud. Use either the Mapboards panel Import option or the Compilation panel's Add Content to add new content. Figure 7.27 shows that a graticule layer has been added to the Compilation panel. This figure also shows the World Countries layer added for reference. Because the World Countries layer is only for visual reference for readers, its syncing is turned off, ensuring that it will not be added to an AI file during a map sync.

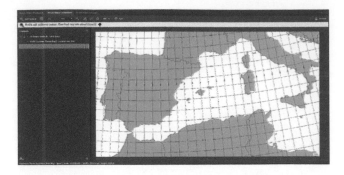

Figure 7.27. A graticule layer from ArcGIS Online is added to the Compilation panel. Because the World Countries layer is only for visual reference, its syncing is turned off, ensuring that it will not be added to an AI file during a map sync. Only the graticule and its labels will be added.

Keep Illustrator layer names generated from map frame names

When new data are synced to a mapboard that was generated from an AIX file, Maps for Adobe Creative Cloud matches the mapboard name to the Illustrator parent layer name that corresponds to the original map frame. For example, syncing the new graticule layer to the Spain Population_Main Map – Spain mapboard will place a new sublayer in the Main Map – Spain parent layer in Illustrator. Changing a parent layer name generated from an ArcGIS Pro map frame name will cause the sync to fail.

2   Add any necessary data filters, labels, geo-analyses, or other desired updates to the layers on the Compilation panel.

Just as with the extension-direct workflow, you can perform the same geo-analyses, style changes, labeling, and filters on new layers added to maps using Maps for Adobe Creative Cloud.

3   Click Sync to download, and add the new layers to the AI file.

With the layer settings appearing as they do in figure 7.27, in which the graticule layer is the only one with its Sync setting on, only this layer and its labels will be added as sublayers to the Spain Population_Main Map parent layer in Illustrator.

When the sync is complete, the new layers and their labels will appear at the top of the corresponding map frame layer on the Illustrator Layers panel. When the graticule layer was synced to the Spain Population_Main Map – Spain mapboard, two new sublayers were placed above the other sublayers within the Main Map – Spain parent layer (figure 7.28). The top sublayer contains any map elements from the sync, and the sublayer beneath it contains any labels and map features added. This second sublayer also includes a serialized appendage to its name indicating the sync's number.

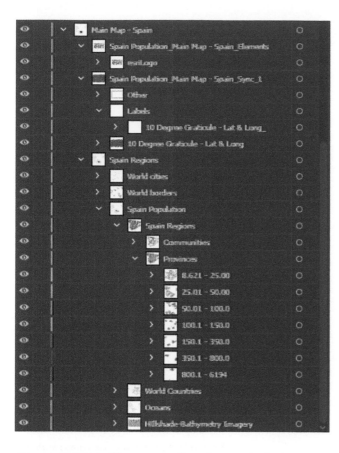

Figure 7.28. In Illustrator, after the graticule layer and its labels are synced, these layers are added as sublayers nested within the Main Map – Spain parent layer.

Rotated maps

If you designed a rotated map and north is not at the top, the mapboard will not display this rotation. However, your map and layers, including layers you add via the Compilation panel, will be correctly rotated in the final AI file.

Additional criteria for adding new data to an AIX file-generated mapboard

If there is a potential that you may add more data to your AI file after exporting an AIX file from ArcGIS Pro, make sure that the respective Export pane follows these criteria:

- If exporting from a map, turn off Clip To Graphics Extent before exporting the AIX file. Doing so will ensure proper data alignment.

- Turn off Enable Wrapping Around The Date Line. This setting is in the ArcGIS Pro Map Properties window on the Coordinate Systems tab.

- When exporting from a layout, make sure that the map frame is entirely within the layout area.

New River Gorge National Park and Preserve map

The ArcGIS Pro–to–Illustrator workflow was used to create the *New River Gorge National Park and Preserve* map. Two map frames were added to a page layout in ArcGIS Pro in the same manner as the Spain population map layout, and they were then used in Illustrator to create the main map and inset map shown in figure 7.29.

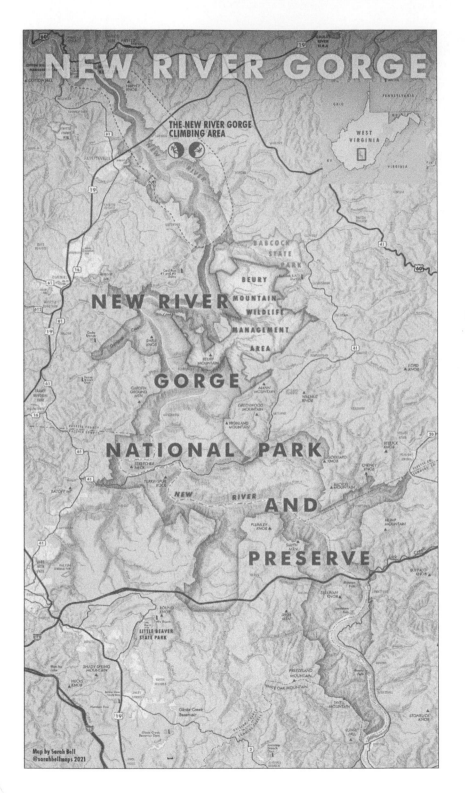

Figure 7.29. I created this *New River Gorge National Park and Preserve* map by following the ArcGIS Pro-to-Illustrator workflow. In ArcGIS Pro, I added two map frames to a page layout—one for the main park map and another for the inset map in the upper-right corner. *Data for this map comes from the USGS, Esri, and the State of Kentucky.*

ABOUT ESRI PRESS

At Esri Press, our mission is to inform, inspire, and teach professionals, students, educators, and the public about GIS by developing print and digital publications. Our goal is to increase the adoption of ArcGIS and to support the vision and brand of Esri. We strive to be the leader in publishing great GIS books and we are dedicated to improving the work and lives of our global community of users, authors, and colleagues.

Acquisitions

Stacy Krieg
Claudia Naber
Alycia Tornetta
Craig Carpenter
Jenefer Shute

Editorial

Carolyn Schatz
Mark Henry
David Oberman

Production

Monica McGregor
Victoria Roberts

Marketing

Mike Livingston
Sasha Gallardo
Beth Bauler

Contributors

Christian Harder
Matt Artz
Keith Mann

Business

Catherine Ortiz
Jon Carter
Jason Childs

For information on Esri Press books and resources, visit our website at esri.com/en-us/esri-press.